GLOBAL BLACK FEMINISMS

This timely and informative volume centers how global Black feminist narratives of care are important to our contemporary theorizing and highlights the transgressive potential of a critical transnational Black feminist pedagogical praxis.

This text not only details how such praxis can be revolutionary for the academy but also provides poignant examples of the student scholarship that can be produced when such pedagogy is applied. Drawing on narratives from Black women around the globe, the book features chapters on pedagogy, mentorship, art, migration, relationships, and how Black women make sense of navigating social and institutional barriers. Readers of the text will benefit from an interdisciplinary, global approach to Black feminisms that centers the narratives and experiences of these women. Readers will also gain knowledge about the historical and contemporary scholarship produced by Black women across the globe.

This book is an invaluable resource for scholars and researchers, including graduate students in Caribbean feminisms, Black feminisms, transnational feminism, sociology, political science, the performing arts, cultural studies, and Caribbean studies.

Andrea N. Baldwin is an Associate professor in the Divisions of Gender and Ethnic Studies in the School for Cultural and Social Transformation at the University of Utah.

Tonya Haynes is a lecturer and Coordinator of Graduate Programmes at the Institute for Gender and Development Studies: Nita Barrow Unit (IGDS:NBU).

Routledge International Studies of Women and Place
Series Editors: Janet Henshall Momsen,
University of California, Davis

Women Divided
Gender, Religion and Politics in Northern Ireland
Rosemary Sales

Women's Lifeworlds
Women's Narratives on Shaping their Realities
Edited by Edith Sizoo

Gender, Planning and Human Rights
Edited by Tovi Fenster

Gender, Ethnicity and Place
Women and Identity in Guyana
Linda Peake and D. Alissa Trotz

Brokering Circular Labour Migration
A Mobile Ethnography of Migrant Care Workers' Journey to Switzerland
Huey Shy Chau

Climate Change, Gender Roles and Hierarchies
Socioeconomic Transformation in an Ethnic Minority Community in Vietnam
Phuong Ha Pham and Donna L. Doane

Intersectional Lives
Chinese Australian Women in White Australia
Alanna Kamp

Bridging Worlds - Building Feminist Geographies
Essays in Honour of Janice Monk
Anindita Datta, Janet Momsen and Ann M. Oberhauser

Global Black Feminisms
Cross Border Collaboration through an Ethics of Care
Edited by Andrea N. Baldwin and Tonya Haynes

The illuminating essays in *Global Black Feminisms* represent a compelling response from a new generation of Caribbean feminist critics to transnational Black feminist thought. This collection represents one of the finest 21st century contributions to Caribbean feminist thought and Caribbean social and political thought, and will be widely celebrated and appreciated.

Aaron Kamugisha,
Professor of Africana Studies, Smith College

Knowledge production – as a collaborative, community-based practice and as a lived experience between students and teachers, mentees, and mentors – is at the heart of this book. A timely centering of global Black–Caribbean feminisms that goes beyond a simple riposte to western–centric feminisms, this book provides a profound exploration of the complexities and liberatory praxes that are necessary for the full recognition of subjectivities that have been historically oppressed, made invisible, and dehumanized. As you read this book, you realize that teaching and scholarship are not tools to be used for the neoliberal promotion of the self within the academic industry. Rather, they are essential for our freedom.

Nathalie Etoke,
Associate Professor of Francophone and Africana Studies
at the Graduate Center, CUNY

Global Black Feminisms

Cross Border Collaboration through an Ethics of Care

Edited by Andrea N. Baldwin and Tonya Haynes

Routledge
Taylor & Francis Group
LONDON AND NEW YORK

First published 2024
by Routledge
4 Park Square, Milton Park, Abingdon, Oxon OX14 4RN

and by Routledge
605 Third Avenue, New York, NY 10158

Routledge is an imprint of the Taylor & Francis Group, an informa business

British Library Cataloguing-in-Publication Data
A catalogue record for this book is available from the British Library

ISBN: 978-0-367-69853-9 (hbk)
ISBN: 978-0-367-69854-6 (pbk)
ISBN: 978-1-003-14355-0 (ebk)

DOI: 10.4324/9781003143550

Typeset in Times New Roman
by SPi Technologies India Pvt Ltd (Straive)

To former, present, and future students.

Contents

Figures

Contributors

Kimberly Nicole Williams is a doctoral student in the English Department at the University of Florida where her work encompasses Black nonbeing, Black healing, and sound studies across multimedia and literature. She has held fellowships with the Callaloo Oxford Residency, the Sweetland Digital Rhetoric Collaborative, and the National Humanities Council. You can find her works in *Sounding Out!*, *Journal of the Society for American Music*, *Peitho*, and more. She is most likely eating blueberry waffles, listening to music, and coveting your dog.

Rachel Afi Quinn is an Associate Professor in the Department of Comparative Cultural Studies and the Women's, Gender & Sexuality Studies Program at the University of Houston. Her transnational feminist cultural studies scholarship focuses on mixed race, gender and sexuality, Black feminist biography, and visual culture in the African diaspora. Her writing on race and culture has been published in *The Black Scholar, Small Axe, Latin American & Latinx Visual Culture, Meridians, Sinister Wisdom*, and *Burlington Contemporary*. She was part of the team that produced the documentary film *Cimarrón Spirit* (2015) on Afro-Dominican cultural traditions. Her first book, *Being La Dominicana: Race and Identity in the Visual Culture of Santo Domingo*, was published in 2021. Quinn was a 2022–23 Scholar-in-Residence at the Schomburg Center for Research in Black Culture for her biographical project on Philippa Schuyler and is recipient of a 2023–24 NEH Faculty Fellow Award.

Maurine Ogonnaya Ogbaa is a PhD Candidate in Literature in the Department of English at the University of Houston. Her research uses feminist theory and gender analysis from social sciences and humanities to analyze West African literature. Her current research project centers theories and structures of development and feminist theories in contemporary coming-of-age novels. Her creative work has been published (or is forthcoming) in *Callaloo*, *Prairie Schooner*, *AGNI*, *third coast*, and elsewhere.

Dr. Daniele Bobb is an Afro-Caribbean scholar whose research and interests include gender and development, government and social policies, mothering, and women and work. Among her literary work is a PhD thesis which is centered on how women negotiate and navigate motherhood and work within the context of neoliberalism. Dr. Bobb completed a BSc in Psychology with Political Science (First Class Honors) and a Master of Philosophy in Political Science at the Cave Hill Campus. She is a lecturer at the IGDS: NBU, the UWI, Cave Hill. In addition to publishing in the area of gender and education, and gender and sexuality, she co-authored the book "Marginalized Groups in the Caribbean: Gender Policy and Society", and has several publications forthcoming. She is involved in many outreach and research projects focusing on a myriad of areas including youth empowerment, the marginalization of vulnerable groups, gender and infrastructure, gender and religion, and gender and education. Dr. Bobb sits on several committees and faculty boards. She is steadfast in her devotion to the work for gender equity and enhancing the quality of life for all.

Leslie Robertson Foncette is a Trinidadian American sociologist whose work focuses on gender inequality and development, agency, and resistance strategies in the Caribbean. She is also a trained bilingual psychotherapist with international experience and holds an MS in Clinical Psychology, a graduate certificate in Women's and Gender Studies, and a PhD in Sociology. Dr. Foncette has been a photographer documenting cultural festivals and rituals, particularly Black life in the Diaspora and Trinidad and Tobago carnival. She integrates the visual lens with a sociocultural exploration of traditions and resistance to oppression. Her scholarship includes exploration of the modes of resistance in the rhetoric and performance of Trinidad and Tobago carnival; social inequality and school violence in Trinidad and Tobago; and the sexual agency of adolescent girls. Dr. Robertson Foncette is currently a SUNY PRODiG Fellow and Visiting Assistant Professor at SUNY Empire State University.

Andreza Jorge is a PhD student in the interdisciplinary Humanities ASPECT program at Virginia Tech University where her research is focused on Cultural and Social Studies. She is also a PhD student in the PPGAC/Performing Arts program at the Federal University of Rio de Janeiro. Andreza's research interests include examining the relationship between Black women and the favela, the educational processes in social projects, the construction of Black and favela women's self-esteem, the method of identity recognition, Black Feminism(s), Afro-Latin dance, performance, art, and Afro-Latin women's agency.

Dr. Nana Afua Yeboaa Brantuo is an interdisciplinary social scientist, policy analyst, and mixed methods researcher with a decade's worth of experience in research, philanthropy, nonprofit, and higher education institutions, along with an extensive track record of policy research and advocacy, teaching and facilitation, and organizational project management and capacity-building. Her work focuses on socioeconomic mobility and immobility, social policy, and civic engagement and advocacy (of Black immigrant communities specifically). She is the founder of Diaspora Praxis, LLC, a research consultancy, where she builds on years of research, writing, and advocacy experience to develop research-driven products that center qualitative and quantitative analysis of socioeconomic, civic, legislative, demographic, and economic data, across time and borders.

Barby Asante is a London-based artist, educator and researcher. Asante applies a decolonial Black feminist lens, to her practice-based research drawing on Akan Adinkra symbology and philosophical principles to consider ways to undo the persistent legacies of slavery and coloniality. She received her PhD in 2022 from CREAM (Centre for Research in Education, Arts and Media) at the University of Westminster in London. She has been a visiting lecturer in several institutions in the UK and the Netherlands and is currently a lecturer in Fine Art Critical Studies at Goldsmiths College in London.

Alexandra Chandra is an experienced marketing professional with a successful track record managing accounts from LGBTQ+ starts ups to large online global brands. Her identity as a South Asian trans woman defines her passion for building platforms that support everyone. She combines creative content and strategy to make culturally relevant work. Through a strategic, progressive approach, she connects content and social justice to build community in a divided world. Alexandra is currently situated on Cahuilla territory – presently Palm Springs California.

Evette Burke has a background in teaching, research and program development for women and girls. She completed her Sociology degree in 1997, postgraduate program in Development Studies in 2003, and her PhD in Gender and Development Studies in 2018. Evette has worked as a Technical Coherence Officer for a UN/EU project that focused on women. She has coordinated the Women's Studies Unit of the University of Guyana as well as taught Research Methods and Gender and Development Studies. She headed a women's desk at the Adult Education Association and worked with the Christian Children's Fund of Great Britain/Guyana. She has also developed, coordinated, and facilitated several training programs for UN agencies in Guyana. Evette is interested in gender equality and prison reform.

Foreword

In this edited collection, Andrea N. Baldwin and Tonya Haynes have co-created a brilliant body of work on the transformative potential of global Black feminist theorizing and praxis. The editors chose contributors that bring regional, transnational, and international perspectives to collectively center, "our black feminist selves as well as other black women globally and their experiences" (Baldwin and Haynes Introduction). This centering is compelling and extremely productive. The authors effectively utilize their diverse locations and theoretical reflections to unearth the emancipatory potential of this simple but profound act of theorizing location initiated through self-reflection.

Their centering of selves begins with a problematizing of geospatial dimensions and a theorization of locations; their sites of working, living, loving, healing, and being. This centering is not about giving prominence to another group of scholars-activists-artists eager to stake their feminist knowledge claims, necessary as these claims are. So, whether these locations include Caribbean or North American physical or virtual classrooms, seminars, or faculty lounges, British or European performance arenas, African\Caribbean migrant networks, home, work and community spaces, Brazilian favelas, Guyanese youth centers or Trinidad and Tobago public health clinics, the authors problematize the politics, power plays, and performance within space.

In the process, they deliberately expose embedded hierarchies of power, of race, of class, of gender and the multiple intersections of these polarities of differing forms of power that reproduce denials and exclusions, whether of race, of class, of gender, or of sexualities. They do this while acknowledging the types of power they themselves may hold and can wield in many, but not all of these contested sites. Throughout the text, spatiality is theorized – promoting and protecting safe and nurturing spaces, creating fugitive spaces, rejecting racist, oppressive spaces. It is striking the theoretical openings and emancipatory practices that emanate from their refusal to take spaces, locations, and sites for granted.

The writers demonstrate that at the core of every contestation, whether it arises in racist ideologies, theoretical hegemonies, academic imperialism, deficient state policies, denigrated sexualities, or the containment of a range of gender identities and sexual pleasures, that foundational to these tensions, is a struggle over power. Race, class, labor, the academy, sexualities, and migration are circulating, overlapping sites where asymmetries of power are enacted to reinforce racism, androcentricity, hierarchies, denials, misogynoir, marginalizations, and exclusions. Collectively these writers expose how these various forms of injustices congeal to maintain the privileges and punishments of heteronormativity and other embedded injustices.

What is also satisfying to discern in these chapters is how these scholar-activists, performance artists, poets, rebel women, queer theorists, transgender women and men, are collectively rupturing existing categorizations and normative orders. They deploy a variety of theoretical frameworks, methodologies, and conceptual tools to reject imposed theoretical and material marginalizations and spatial exclusions. In the process, they continue to circumnavigate geographic and psychic borders. They are dedicated to remapping geographies of place and time. Yet, they are continuously evaluating, rethinking, and living, inhabiting fully all the sites and spaces in which they experience the fulsomeness of being Black, insurgent global feminists and activists.

As a result, this text is particularly beneficial for persons who inhabit virtual or physical pedagogical spaces as professors or students, even as it demands that one critiques the traditional positions of power that coalesce asymmetrically around those who inhabit such locations.

While Baldwin, Haynes and the contributors emphasize temporality, geo-spatiality and the contemporary contributions of global Black feminist scholar-activists, they ensure they identify the feminist genealogies that brought them to this moment, that enable them to encircle Black feminist scholarship coming from Africa, African diasporas, African and Asian Americans, South American and Caribbean feminist scholars.

These chapters move beyond naming and interrogating contested locations to investing in empowering practices of mutual mentoring. This shared strategy constitutes a unique approach to enacting an "ethics of care". Toward this end, the collection invests heavily in theorizing pedagogical practices to reimagine classroom spaces as less hierarchical and a site for genuine sharing and mutual mentoring. The authors reveal heavy intellectual and practical investments in mentorship and theorizing the process of mentoring as a fulsome expression of the ethics of care.

In this text, while engaging with existing theorizing, the authors contribute original feminist theoretical frameworks and feminist methodologies as part of an overarching commitment to developing and enacting a decolonial knowledge project. They do this by exploring exciting thematic areas inclusive of:

- Transactional belonging;
- Enacting a decolonial feminist project;
- Recentering and expanding the epistemological scope of women studies;
- Mothering across class structures in the absence of supportive state policies;
- Feminist performance politics and poetics;
- Spirituality and healing in LGBTQIA+ communities;

- Theorizing an ethic of care in the academy;
- African diasporic migrations;
- Feminist problematizing of pedagogical sites and practices;
- Confronting powerlessness in state institutions;
- African philosophic traditions.

The emerging theoretical frameworks they offer include "Black Favela Feminisms", and "Critical Transnational Queer Praxis and Theory". The writers generate a range of original feminist conceptual tools that are even more satisfying, since these concepts are accompanied by very useful definitions:

- Tacit sexualities;
- Homolocalism;
- Masturbating methodologies;
- Psychosocial uncertainty;
- An ethics of care in pedagogical practices.

As I read the collection, I kept returning to the idea that these offerings are contemporary expressions of the ruptures and feminist visioning that Lucille Mair's pathbreaking dissertation, "A Historical Study of Women in Jamaica 1655–1844", signaled in 1974. Baldwin and Haynes declare that in their edited collection, this "text is a specific type of global Black feminisms – Caribbean feminisms" (Baldwin and Haynes Introduction). Their insistence on ensuring that their latest contribution to Caribbean feminist thought remains within the canon of global Black feminisms echoes the feminist intellectual thrust of Mair's pivotal study.

Sometime ago, I assessed the intellectual thrust of Mair's scholarship as constituting ruptures and visionings in Caribbean feminist theorizing and as laying the groundwork for indigenous feminist epistemologies. I stated "by challenging the then constructed invisibility of the Black\African\Caribbean woman during slavery, Mair recentered the subjectivity of the Caribbean woman and provided critical openings for future feminist theorizing" (Barriteau n.d. [2003]). Forty-nine years after Mair's study, *Global Black Feminisms: Cross Border Collaborations through an Ethics of Care* has not only transcended the critical epistemic openings Mair's thesis anticipated, but it has broken new ground, constructed new theoretical frames and analytical tools while illuminating previously uncharted pathways for ongoing engagements.

Tonya Haynes ensures that critical engagements with the works of Sylvia Wynter continue to generate fertile ground for feminism's contemporary. In Chapter 1, "Women's Studies After Wynter", Haynes theorizes Sylvia Wynter's epistemic works as being very disruptive to feminist orthodoxy on gender even while acknowledging a reluctance or ambivalence among feminists to engage with Wynter's theorization. Haynes states that Wynter's work suggests a reclaiming of the emancipatory potential of the epistemic rupture signaled by the emergence of women's studies. She declares that "Wynter challenges both the hegemony of gender within feminist theorizing and how gender is defined, questioning the foundational assertation that 'woman' is Man's other".

Black feminists and activists within and beyond the binding, blinding boundaries of the academy can learn from what these writers have gifted to contemporary global feminist analyses and praxis. In Chapter 11, "University Plantation Il/logics: Black Women's Fugitivity and Futurity in the Wake of COVID-19 and the Global Anti-Racist Uprisings of 2020/21", Baldwin asserts that she is following Katherine McKittrick's position that she can learn from those who came before. It is clear these theorists\activists\artists have invested in interrogating the works of Sylvia Wynter, Lucille Mair, Ama Ada Aidoo, Jacqui Alexander, Audre Lorde, and so many others. More critically, they are charting original, emancipatory, intellectual, and activist projects. In the process, they have honored the philosophy of Sankofa, the Twi\Akan\Ghanaian concept of retrieving the best from the past to carve more productive and meaningful futures. This philosophy permeates the text and informs critical dimensions of theorizing and activism of several of the writers.

In these eleven chapters, the authors invoke the scholarship, activism, praxis, and subversive theorizing and living of those who went before. In the process, these writers have expertly interrogated the abundance of the highly productive work they have inherited. They have formulated new theoretical frames, articulated and expanded feminist conceptual vocabularies, problematized and devised new methodologies for recentering Black feminists' ontologies. They continue to carve indigenous feminist epistemologies for which *Global Black Feminisms* … is a compelling, contemporary expression. Andrea Baldwin and Tonya Haynes offer their work and that of their contributors to situate Caribbean feminist thought and praxis among an organic intellectual community of global Black feminisms. They have succeeded.

Eudine Barriteau
January 29, 2023

Reference

Barriteau, Eudine. (n.d. [2003]) 'Report on the Research Project Ruptures and Visioning: Lucille Mathurin Mair and the Emergence of a Caribbean Feminist Epistemology'. Centre for Gender and Development Studies, University of the West Indies, Cave Hill Campus.

Acknowledgements

We would like to acknowledge our colleagues at The University of the West Indies, especially Professor Aaron Kamugisha and Dr. Debra Providence, for their intellectual generosity and support and Dr. Angelique Nixon for her feminist friendship. We express our sincere gratitude to colleagues of the Institute for Gender and Development Studies: Nita Barrow Unit: Professor Eudine Barriteau, who continues to be a generous mentor, and Dr. Halimah DeShong, a treasured friend as well as Dr. Daniele Bobb, Leigh-Ann Worrell and Veronica Jones.

We would also like to acknowledge the generous support from Professor Julia S. Jordan-Zachery and Professor Nathalie Etoke. Many thanks to all who reviewed the chapters in the text, thank you for your time and extensive feedback.

It goes without saying that this text would not be possible without the groundbreaking scholarship of the authors whose work is featured herein. Their commitment to global feminist scholarship and this text is a clear indication that the future of this work is on solid footing. To them we are extremely grateful!

Working on this collection coincided with one of those most heart-breaking times of my life and a period of crisis for my family. I thank my family, colleagues, and friends for their presence. I have to give a special shoutout to my niece, Kyalh Williams, who lights up our lives with her fierce stubbornness and quick wit. Alitia Quintyne, Alana Cadogan, Sherry-Ann Haynes, Dottine Haynes, Angelique Nixon, Halimah DeShong, Aaron Kamugisha, and Tami Navarro, thank you so much for checking in and checking up on me even as you weathered your own storms (Tonya).

I am grateful to my support network including my mother Sandra Griffith, my husband Quentin T. Baldwin, and son Liam T. Baldwin. As always, a special thank you to Dr. Nana Afua Yeboaa Brantuo for her continued friendship, collaboration, and support. I am also eternally grateful to Professor Jordan-Zachery for the extraordinary amounts of care she has extended to me and my family at the time of COVID and during the tumultuous year of career transitions, including the tenure process, job search, interviewing process, job negotiations, and the ways she has constantly held my family and in much closeness (Andrea).

We recognize all of the revolutionary Caribbean women with their complex and complicated relationship to feminism whose praxis of care has made our work possible.

Home-Grown and Grounded

Black Caribbean Feminist Pedagogies in Global Conversation

Andrea N. Baldwin and Tonya Haynes

Introduction

At the 2022 graduation ceremony of the Cave Hill Campus of the University of the West Indies, honorary graduate, Ambassador Gabriel Abed, advised students that the Caribbean was just a sandbox where they could get their start: "Another important lesson I learnt is, go global. Barbados is a very ideal sandbox. It's a great place to start off. You shouldn't be bound by these 166 square miles. Your value is simply too big for Barbados." We experienced this advice as equal parts ignorance, insult and irony. A sandbox, according to the Oxford dictionary definitions, can either mean "a shallow box or hollow in the ground partly filled with sand for children to play in," or in the computer science field, a testing environment.[1] Both of these definitions conjure up Barbados, and the Caribbean in general, as a space which limits the potential of our people, a space so constrained that it is only good for a practice run, and a test of what we are truly capable of – a shallow play place. *A small place*. But there is no sandbox in our childhoods. There was the beach, but the real treat was the sea bath, not the sand. A bath, not a swim, because our parents were afraid the sea would swallow us alive as it had done so many. *De sea engot nuh back door*. No way of slipping out in an emergency if you got into trouble. We had only encountered the sandbox via US TV as a place for white children to play in. These repeating islands and territories – "the intimacies of four continents" (Lowe 2015) – will perhaps always remain a sandbox in the colonized tourist imagination of so many. Dionne Brand writes in *A Map to the Door of No Return* that, "Through the BBC broadcasts we were inhabited by British consciousness" (2002, 16–17). Through globalized media there is an attempt to socialize us all into a neoliberal entrepreneurship of the self, but all over the world, people are fighting back.

In her interview with Bedour Alagraa (2021), Sylvia Wynter affirms that her creative work with *Jamaica Journal* informed her praxis of "enacting an autonomous space for transformation":

> When you ask me what the universities thought about Black studies back then, I will tell you that nobody even thought about it! (*laughs*). It's that

DOI: 10.4324/9781003143550-1

> simple. We were starting from a *negation* on the part of the institution, but I had come with all the knowledge and language I had developed on the creative side of the *Jamaica Journal*! I did not enter these conversations concerning Black studies with the institution in mind, but rather with the *Jamaica Journal's* practice of enacting an autonomous space for transformation, and I had always viewed Black studies as a way of trying to enact a transformation of knowledge, and we did go pretty far at it!

The radical potential of Caribbean creative practices to undo "the stereotyped view of yourself that you yourself have been socialized to accept," to displace and disrupt the sea-sun-sand-and-sex sandbox stereotype and its attendant imperialism, anti-Blackness and misogynist repertoires, is for us, at the center of our commitment to Black Caribbean feminist knowledge production (Wynter in Scott 2000, 131).

In 2007 we began our course of study for the doctoral degree in Gender and Development Studies as part of the first cohort of students at the Institute for Gender and Development Studies, Nita Barrow Unit (NBU) at the University of the West Indies Cave Hill. I (Tonya) received a letter by post in 2006 informing me that the NBU would be offering Master and Doctor of Philosophy programs in 2007. I didn't know at the time that the NBU's full-time teaching staff were only two, inclusive of Professor Barriteau, the Head of the NBU, promoted to full professor in 2004, or that Professor Barriteau would spend four years simultaneously as Head of the NBU and Director of the School for Graduate Studies and Research. (No one else has held these double appointments since as both are considered separate full-time roles.) In 2006 I (Andrea) had no idea who Professor Barriteau was. In my course of study up to that point I had developed an interest in examining gender as it related to the law and trade. It was another professor with whom I had worked as part of my master's degree who told me about the new program at NBU and suggested to meet Professor Barriteau. We both had no sense of the labor and love that went into this vision of creating a new generation of home-grown scholars. We were unaware of the history of the Institute for Gender and Development Studies, the pathbreaking work of the Women and Development Studies groups which preceded it, and the seemingly superhuman work ethic which feminist institution-building demanded. We were just excited to make our way to class in the evening, after, for Tonya, a long, stressful day of teaching in a secondary school, and for Andrea, a demoralizing day in the law courts and then interfacing with international organizations touting Barbados's value proposition. We came to know intimately the care with which Professor Barriteau treated every single student, her exacting high standards, and like everyone else in that cohort, rose to meet these expectations.

Of course, you don't simply run a doctoral program with two members of the teaching staff. A transnational community of Caribbean feminist scholars and activists committed to love as a rule provided support to the program. For example, Patricia Mohammed, based in Trinidad and Tobago, supervised Tonya's doctoral thesis. What an education at the University of the West Indies provided was a grounding in Caribbean history, languages and anti-colonial feminist thought. To make sense of the Caribbean and our place as Black women within and outside of it, you necessarily have to make sense of the world.

Gloria Knight, cited in Mary Chamberlain's essay on historian Elsa Goveia states,

> I went into a lecture room one day when Elsa Goveia was lecturing and I was so turned-on, I couldn't believe that history could be like that... It is something that I will not forget. She was a wonderful teacher. I never missed a lecture by her, it was a new world opening before me every time.
>
> (2004, 182)

Our experience in the undergraduate classroom at Cave Hill with Eudine Barriteau, Tracy Robinson and Michelle Rowley and in the graduate courses led by Eudine Barriteau, Halimah DeShong and Aaron Kamugisha was the same – *a new world opening before us every time*. Michelle Rowley writes that "a feminist pedagogy that is able to incite our students to riot is a pedagogy that also incites our students to love" (2007, 145). It is that kind of erotic feminist pedagogy which grounds this text and the ethics of care which produced it. We too are part of Elsa Goveia's legacy. Elsa Goveia was the first woman to achieve the rank of full professor at the University of the West Indies and she supervised Lucille Mathurin Mair's 1974 doctoral thesis, "A Historical Study of Women in Jamaica 1655–1844." This pioneering work was completed nearly 20 years before the establishment of the Centre for Gender and Development Studies.

In August 2020 the auditorium of the Ellerslie Secondary School, Professor Barriteau's alma mater, was renamed the V. Eudine Barriteau Auditorium.

> If those two teachers hadn't arranged for a transfer for me at 13 plus [from St. Matthias All Age School to the recently built Ellerslie Comprehensive School], I would have left school at 14! ... The thing is I would not have known what I was missing!

Professor Barriteau stated. Professor Barriteau also noted the quiet activism of women teachers whose efforts kept girls in school beyond the school leaving age before official policy increased it to 16. This expanded education access brought about by reforms of the 1960s and 1970s is fundamental to her own intellectual and leadership trajectory – a career in Higher Education dedicated to the Caribbean. Her institution-building work cleared a path for the entry of research students in the field of Gender and Development. For Caribbean people with a history of colonization, enslavement, indenture and indigenous genocide as well as enduring ableist and classist exclusions from education, access to higher education cannot be taken for granted, and our responsibility to the community and the wider world remains a debt that is never settled.

Our experiences in the Gender and Development Studies program over a decade ago taught us how to be feminist researchers, pedagogues, community advocates and mentors. And as we have engaged in this work over the years it has become clear the value of centering these experiences as we also pivot (Collins 2000), recognizing them as partial as we seek to engage in transnational feminist work. Particularly in the last few years as we have been grappling globally with the impacts of the COVID-19 pandemic and the disproportionate and inequitable response, such that African and Caribbean nations have had a difficult time accessing vaccines from global supply stores, and Black and Brown people in the West are finding it difficult to access care even when these wealthy nations have continued to hoard vaccines. In addition, the impacts of climate change have meant that the most vulnerable people are those who emit the least CO_2 gasses, and small island states, and indigenous communities are feeling the brunt of a warming climate. Add to that, inflation in the West and the war in Ukraine have impacted global supply chains and the cost of food and other essentials globally with no end in sight. Although we are all impacted differently by crisis after crisis, what we can be certain of is that we are all impacted. What is more, because of global anti-Black racism and a system of patriarchy, Black women historically have borne the brunt of whatever crisis the world faces.

But we the editors, as we recognize these global hardships for Black women, also know, we are part of a feminist lineage which has never "shied away from …[a] vision of liberation [which] is arguably the most radical twenty-first century vision articulated in the region" (Haynes 2012, 63). So, as we center our experiences during these multiple crises, it is essential that we not only recognize how Black women across the globe are coping in our contemporary moment but that we also pay attention to how they/we are making sense of it. As such, this text is an exercise in centering – our Black Caribbean feminist selves as well as other Black women globally and their experiences – and pivoting. Each chapter centers the author's unique experiences and together these experiences provide us with a more expansive view of how Black women globally are making sense of themselves and this world. The authors in this text come from various places around the globe including the Caribbean, Latin America, the African continent, the United States and Europe. Each of these Black women contributors has something to tell us about what it is to live as a Black woman in this contemporary moment, the issues which are most pressing for them, the uncertainties they experience daily, the strategies they are engaging to combat daily oppression, and how they are being affected by these. By simultaneously centering the experiences of these Black women, the text allows the reader to pivot from one experience to the other, finding commonalities and differences. We echo Rachel Afi Quinn and Maurine Ogonnaya

Ogbaa, who in their chapter for this text quote Rodriguez, Tsikata and Ampofo (2016): "we affirm and celebrate the idea that black feminist scholars and those who seek to understand the lives of women in Africa and the diaspora have multiple identities and choose to speak in many voices" (ix).

Prior to the start of the pandemic, this text was conceived as one which centers how teaching and mentorship can be a transformative tool for Black women scholars; as a text attentive to how other Black women globally, particularly graduate students, engage in and experience *a new world opening before them every time*, like we had. However as with most projects planned just prior to the start of the pandemic, we had to do a lot of pivoting as there were many unforeseen circumstances which occurred. Folks got ill or took on the care of friends and family who became ill. It was difficult to find reviewers because of burn out. Folks took up new positions, and in general we were all just really exhausted. When the majority of the contributors submitted draft essays to this volume over a year ago, they were at some stage in their graduate school journey. Now as we go to press, we are happy to write that many of them have since graduated or are on the cusp of completing their graduate school journey. When the editors solicited essays from the authors – students who we have personally taught and mentored, or those who have been taught or mentored by colleagues, as the chapters by Maurine Ogonnaya Ogbaa and Rachel Afi Quinn, and Andrea N. Baldwin and Alexandra Chandra in this volume exemplify – we were building from our own experiences in our doctoral program from over ten years ago. Modeled for us was a pedagogy which stems from a belief in the possibility of "education as the practice of freedom" (hooks 1994, 207). This pedagogical praxis we know full well, unsettles, imagines and enacts an otherwise. It is a practice of self-creation and (re)naming. It is based on an ethic of care (Baldwin 2021) and love; similar to that which brought Caribbean feminists from around the globe together over a fifteen years ago to realize a vision of building the Gender and Development Studies program.

The legacy of this program lives on in this text not only in the editors but through the work of Daniele Bobb and Evette Burke, who, like us, are graduates of the NBU, Evette was supervised by Professor Barriteau, Daniele by Tonya. This legacy also includes Leslie Robertson Foncette who Andrea advised through her graduate study at Virginia Tech. Their chapters "Mothering in Neo-liberal Contexts: Caribbean Women's Experiences," "Tacit Sexualities: Transforming the Narrative" and "Psychosocial Uncertainty: Making Sense of Institutional Suffering in Trinidad & Tobago" all theorize Caribbean women's lives. Bobb examines how Caribbean women negotiate motherhood and work in the context of our neoliberal Caribbean societies, concluding that where Caribbean mothers resist the hegemonic institution of motherhood they do so through neoliberal discourses of work. This resistance is based on the inherent power relations of gender existing in the region, which arise from the categorized expectations of family, friends and normalized beliefs, that often leave the mother with feelings of incompetence and inadequacy. In her chapter, Burke details interviews with non-marital, non-cohabiting Afro-Guyanese women who engage in what she calls "tactic sexuality," that is sexual relations "in which sex, functions in coterminous agreement/value with emotional investment and financial social support, to sustain the relationship," to complicate the ways that Afro-Caribbean women's sexuality is perceived and theorized as simply transactional. Burke argues that in these relationships of tacit sexuality "the affective attachment creates increased familiarity which allows for the performance of non-sexual activities. These relations of high social tolerance tend to be fractious because of the often-unstable commitment to love, sex, and money on which they rest." Foncette in her chapter details the ways in which Trinidadian women make sense of navigating maternal and reproductive care. She draws from personal experience as a Trinidadian woman, incidents described in news media, and accounts of people living in Trinidad and Tobago to explore what happens when people meet challenges or misfortune engaging with everyday public services. Bobb's, Burke's and Foncette's thoughtful theorizing about Caribbean women's lives would not be possible at all had it not been for the NBU program.

The program cleared the way for this work to exist. It was/is a clearing in the way that Kimberly Nicole Williams in her chapter "The Geography of Healing at the End of the World: Black Scholar Practitioners Who Evoke Toni Morrison's The Clearing" writes about the power and potential of the Clearing created by Morrison's character Baby Suggs in *Beloved* (1987) to think about how Black people create space to enact healing, to share knowledge, to celebrate life, find joy and build community. The clearing in Morrison's text is described as "a wide-open place cut deep in the woods nobody knew for what at the end of a path known only to deer and whoever cleared the land in the first place" (88). It is a place we get to by a cut path, a pathway, cleared for us and imagined by Williams to exist in the physical and communal sites, artistic, spiritual and advocacy-based; like the ones curated by those she interviewed for her essay – Nia Wilson of SpiritHouse, Roc of Tampa's Rooted Resistance, and Alexis Pauline Gumbs who curates Black feminist Sunday Service. Like Williams, we examine how these contemporary sites, including our classroom, are places of the Clearing; Black feminist care geographies.

This text, intentionally titled *Global Black Feminisms: Cross Border Collaboration through an Ethics of Care*, is clear about the criticality of understanding the socially constructed nature of geographical space (McKittrick 2006). The authors in this text attend to the importance of space through attention in their work to the historically and contemporary implications of socially constructed national borders for African diasporic peoples, as with Nana Afua Yeboaa Brantuo chapter "Subversive Knowledges and Praxes of Black Immigrants in the United States: Reflections from a Scholar-Advocate;" and Andreza Jorge's chapter "Black Favela Feminism: The Struggle for Survival as a Transformative Praxis". Brantuo in her chapter honors the geographical gymnastics which immigrants from Africa and the Caribbean at various times in history have had to perform as they crossed borders. In her chapter, she lovingly centers the cultural, intellectual and political traditions and productions of Black immigrant advocates, past and present. Jorge, in her theorizing of the spatio-temporality of Brazil, and of the traditions of Black women who draw orixá from the Yoruba African pantheon called Oiá/Iansã, presents the experiences and narratives of Black faveladas who participated in a social and community project developed in the Complexo de favelas of Maré/Brazil: Mulheres ao Vento (Women of the Wind) project. In her chapter, she coins the term Black favela feminism to indicate the ways in which these women make sense of their lived experiences dealing with racism, sexism and classism and how their counter-hegemonic narratives and performances can inform Black feminisms and help foster social transformations.

The stories relayed by Brantuo and Jorge are an unapologetic record that our "lived experience of theorizing is fundamentally linked to processes of self-recovery, of collective liberation, [such that] no gap exists between theory and practice" but that there is a "reciprocal process wherein one enables the other" (hooks 1994, 61). Their work documents a sustained approach by Black women globally – whether they identify as Black feminists, Caribbean feminists, transnational feminists, as all, some, or none of the above – to doing scholarship otherwise. It is an approach that resists the general tendency in the academy to categorize, isolate and produce competition through discipline, and that is made more pronounced by "a celebrity culture where an internalized need to present oneself as an individual academic star often translates into a drive to abstract and generalize, frequently in opposition to those who are seen as immersed in grounded struggles" (Lock Swarr and Nagar 2010, 2). In their chapter "Public Scholarship as B(l)ack Talk: African Feminist Collaborations in the Academy and Online" about working otherwise in the academy, Rachel Afi Quinn and Maurine Ogonnaya Ogbaa demonstrate how theory and practice are fundamentally linked as they situate their mentorship relationship as an invaluable resource and feminist practice. They center African feminist thought in their respective work, while also pivoting to engage African diasporic, African American and transnational feminist thought as theoretical frameworks they deploy in their scholarship. They explain how this theorizing and collaborative work across disciplines – in public scholarship in particular – allows them to carve out spaces for their research in and beyond the academy as well as reimagine an alternative future for themselves as they contend with unique forms of isolation and institutionalized misogynoir. Similarly, in their chapter "Critical Transnational Queer Praxis: Perspectives on (Re)Production, Performance, and Punishment in the Academy" Andrea N. Baldwin and Alexandra Chandra write about developing what they call a critical transnational queer praxis grounded in their experiences navigating the academy. They describe that central to this praxis is queer and Black and transnational feminist theorizing, cross-border community building and collaboration. The praxis pulls and gestures toward radical genealogies past and present to re-spatialize geographies of power in order to create spaces in the academy that reach beyond borders to enact minoritarian liberatory performances. Performances much like Barby Asante's Declaration of Independence detailed in "The

Women. They Were Plotting Too: Declaring our Independence in the Spirit of Sankofa." Asante details in this chapter the intentionality with which she created and curated the Declaration of Independence, a performance in which womxn "come together in a powerful and emotive presentation of ourselves, bringing with us our anger, our grief, and our celebration in a ritual circle that is as much a resource for us as it is an artwork." According to Asante, *Declaration of Independence* is a refusal to be complicit but rather uses otherwise and other ways of knowing and understanding, such as the Akan Adinkra principle of *Sankofa* loosely translated from the Akan languages of Twi and Fanté as "retrieve" or "go back and get."

The aforementioned chapters demonstrate how critical good mentoring relationships are for Black and Brown (cis and trans) women in the academy. It is no coincidence that our own teacher and mentor Eudine Barriteau has written the foreword for this text. And it is with intentionality that we write this introduction also as a tribute to her and other Caribbean feminist scholars mentioned here by name and others who are not. We celebrate those whom we knew from jump and those who like Professor Julia Jordan-Zachary, who has written the afterword for this text, we came to know and love later along this journey. We honor them here for teaching us how to do our "own dreaming, … own thinking, and … own planning" (Brodber 1990, 166), and we hope that the spirit of intentionality, recognition and appreciation demonstrated in this volume emanates beyond its pages. We hope that the Black women in this text can become a community from whose voices, stories and knowledges readers can draw on to challenge, inform, teach and guide them along their own individual and collective journeys. In this vein we envision this text as part of our own – editors and authors – pedagogy and teaching practice of love which reflects the liberating potential of teaching, learning and knowledge creation.

We are two Black Bajan,[2] Caribbean feminist pedagogues who shared space as students over ten years ago and who now work across geo-spatial borders – Andrea at the University of Utah in Salt Lake City in the United States and Tonya at the University of the West Indies Cave Hill in Barbados – as friends and comrades. We have divided this text into four thematic sections: 1) Black Feminisms: Sites of Black Feminist Existence, 2) Black Women's Lived Experiences in Our Contemporary Societies, 3) Black Feminist Activism: A Worldmaking Praxis of Care, and 4) Black Feminisms and Healing Futures. Our chapters bookend the others in these sections and are intentionally nestled between the gorgeous afterword and foreword of two Black Bajan women who signal therein the care they have for Black women, and Black and Caribbean feminist theorizing, the care with which this text was curated. They signal that this text is a legacy and lineage of Caribbean feminisms.

Tonya's opening chapter in Section I "Women's Studies after Wynter" is a fitting opening as it blends reflections on theories and teaching of gender with Wynter's unsettled relationship to feminism and gender theory. Andrea concludes the text with her chapter on fugitivity and futurity entitled "University

Plantation Il/logics: Black Women's Fugitivity and Futurity in the Wake of COVID-19 and the Global Anti-Racist Uprisings of 2020/21." In it she acknowledges that even as Black women scholars globally have created and curated space within the contemporary Western neoliberal university, the space continues to be one of trauma and violence, exacerbated during the COVID-19 pandemic and by the outright targeting and killing of Black people in the United States. Building on McKittrick's (2013) theorizing, she argues that the technologies of violence and logics of the plantation move through time and are embraced by the university, even as it uses the rhetoric of diversity, equity and inclusion to disavow this violence. What has resulted in the past few years is an exodus of Black women literally and figuratively fleeing for their very lives. Baldwin reads these fugitivities, whether these Black women are completely leaving the academy or finding temporary but sustained solace and community outside of its structures, as activating and building on the legacies and strategies of Black fugitives from slavery, and death to secure their own and others futurity in this contemporary moment of violence.

The NBU is our wide-open expanse right there in the region, home-grown, spacious, intimate, and deep. Our Black Caribbean feminist selves are expansive, the size of the region and the size of our program do not mean containment! No sandbox here! We are not insulated; we are not held back. We will not buy into this narrative, rather, as Haynes writes, we refuse to be managed as we continue this Caribbean feminist project, one of "continuity and rupture, critical engagement and departure" (2012, 66). What we have put together in this volume is a text which simultaneously showcases the possibilities of affective teaching and mentorship, the importance of cross-border collaboration, the rigorous application of Caribbean, transnational and Black feminist theory to global issues, as well as the brilliant individual scholarship of Ph.D. (and former Ph.D.) students who are molded through this application of critical pedagogy and in whose hands the future of the academy lies. Black feminist anthropologist Irma McClaurin (2001) writes about the ways in which Black women scholars can work to undertake the difficult task of creating decolonial and transformative activist scholarship in order to alleviate conditions of oppression. This, and their work is "a poetics, a practice, that unsettles, disorients, imagines otherwise possibility" (Crawley 2018, 11).

Notes

1 https://www.oxfordlearnersdictionaries.com/us/definition/english/sandbox
2 Colloquial for Barbadian.

References

Abed, Gabriel. 2022. https://www.facebook.com/UWITV/videos/1220260421863967

Alagraa, Bedour. 2021. "What Will Be the Cure?: A Conversation with Sylvia Wynter." *OffShoot*. Accessed 8 November 2022. https://offshootjournal.org/what-will-be-the-cure-a-conversation-with-sylvia-wynter/.

Baldwin, Andrea. 2021. *A Decolonial Black Feminist Theory of Reading and Shade: Feeling the University*. New York: Routledge.

Brand, Dionne. 2002. *A Map to the Door of No Return: Notes to Belonging*. Toronto: Vintage Canada.

Brodber, Erna. 1990. "Fiction in the Scientific Procedure." In *Caribbean Women Writers: Essays from the First International Conference*, ed. Selwyn R. Cudjoe, 164–8. Wellesley, MA: Calaloux.

Chamberlain, Mary. 2004. "Elsa Goveia: History and Nation". *History Workshop Journal*, 58: 167–90.

Collins, Patricia. 2000. *Black Feminist Thought: Knowledge, Consciousness, and the Politics of Empowerment*. New York: Routledge.

Crawley, Ashon. 2018. "Introduction to the Academy and What Can Be Done?" *Journal of Critical Ethnic Studies*, 4(1): 4–19.

Haynes, Tonya. 2012. "The Divine and the Demonic: Slyvia Wynter and Caribbean Feminist Thought Revisited." In *Love and Power Caribbean Discourses on Gender*, ed. Eudine Barriteau, 54–71. Kingston: University of the West Indies Press.

Hooks, Bell. 1994. *Teaching to Transgress: Education as the Practice of Freedom*. New York: Routledge.

Lowe, Lisa. 2015. *The Intimacies of Four Continents*. Durham: Duke University Press.

Lock, Swarr Amanda, and Richa Nagar. 2010. "Introduction: Theorizing Transnational Feminist Praxis." In *Critical Transformational Feminist Praxis*, ed. Amanda Lock Swarr and Richa Nagar, 1–20. Albany, NY: State University of New York Press.

McClaurin, Irma. 2001. *Black Feminist Anthropology: Theory, Politics, Praxis and Poetics*. New Brunswick, NJ: Rutgers University Press.

McKittrick, Katherine. 2013. "Plantation Futures." *Small Axe* 17(3), 1–15.

———. 2006. *Demonic Grounds: Black Women and the Cartographies of Struggle*. Minneapolis: Minnesota Press.

Morrison, Toni. 1987. *Beloved.* New York: Randomhouse.

Rodriguez, Cheryl Rene, Dzodzi Tsikata, and Akosua Adomako Ampofo. 2016. "Introduction." In *Transatlantic Feminisms: Women and Gender Studies in Africa and the Diaspora*, eds. Cheryl Rene Rodriguez, Dzodzi Tsikata, and Akosua Adomako Ampofo, vii–xxxii. Lanham: Lexington Books.

Rowley, Michelle. 2007. "Rethinking Interdisciplinarity: Meditations on the Sacred Possibilities of an Erotic Feminist Pedagogy." *Small Axe*, 24: 139–53.

Scott, David. 2000. "The Re-Enchantment of Humanism: An Interview with Sylvia Wynter." *Small Axe*, 8: 119–207.

Thomas, Greg. 2006. "Proud Flesh Inter/Views: Sylvia Wynter." *PROUDFLESH: A New Afrikan Journal of Culture, Politics & Consciousness*, 4. http://www.africaresource.com/proudflesh/issue4/wynter.html.

Section I

Black Feminisms: Sites of Black Feminist Existence

1 Women's Studies after Wynter*

Teaching Gender and Development Studies in a Gender-Conscious Caribbean

Tonya Haynes

Introduction

Joining the conversation on recent challenges to established genealogies of feminist theory, Women's Studies and women's movements, I argue that Wynter's (cited in Thomas 2006) proposition "that it is the genre of Man that causes all the -isms" (55) demands a reorientation of Women's/Gender/Feminist Studies curricula and pedagogy which does not just read her work and that of other decolonial, Black and transnational feminist scholars as serving as response or corrective to dominant Western narratives about feminist activism and theory. Like Méndez and Figueroa (2020), I read Wynter alongside other Black, Caribbean, anti-colonial and transnational feminist scholars, demonstrating points of convergence and coherence. I have argued that Wynter challenges Women's Studies in foundational ways: she destabilizes gender as the primary epistemological terrain of feminist studies by revealing the epistemic and conceptual complicity of Western gender in our current order of knowledge and its attendant racially exclusionary conceptualization of what it means to be human (Haynes 2012). Her work then suggests a reclaiming of the emancipatory potential of the epistemic rupture signaled by the emergence of Women's Studies. I reflect on my own practice of teaching Gender and Development Studies in the Caribbean, highlighting just what this means methodologically and pedagogically. Specifically, I focus on the teaching of feminist theorizing of gender. I share moments from the classroom which reflect student engagement with Wynter's ideas and academic and popular theories of gender.

Wynter as Black Feminist Thinker

As I'm writing this chapter, Black feminist theory has moved from margin to center and *is* feminist theory. It can no longer be read as a critique or corrective to the mainstream. It is, in many ways, the mainstream itself and is the source of the most generative and urgent scholarship today. While Tinsley (2008) was

* I wish to extend sincere thanks to Aaron Kamugisha whose extensive engagement with the ideas expressed here has served to strengthen them. I also express gratitude to the students who continue to challenge, inspire and shape my teaching and scholarship.

DOI: 10.4324/9781003143550-3

concerned that Black queer studies has been "situated as a dazzlingly new 'discovery' in academia – a hybrid, mermaidlike imagination that has yet to find its land legs" (193), I am afraid that the current Black feminist "moment" may mean appropriation and extraction with little concern for the lives of Black feminist scholars or Black people.

Sylvia Wynter is firmly cemented as a Black feminist scholar. As Alex Weheliye (2014) demonstrates, her theorizing of the human finds resonance with the foundational work of Hortense Spillers. Greg Thomas (2007) and McKittrick (2006) provided some of the earliest and most generative engagements with her work around sexuality and Black women's demonic grounds. More recently, numerous Black feminist scholars have taken up her work to theorize the human, critique post-humanism, advance transaesthetics, Black Studies and Native Studies and a Black and decolonial feminist critique of dominant conceptualizations of gender and sexuality (see Bey 2022a, Jackson 2020, King 2019, Snorton 2017, Stallings 2017). These Black feminist scholars take seriously Wynter's assertion that:

> Black women's struggle is quite other. Our struggle as Black women has to do with the destruction of the genre; with the displacement of the genre of the human of "Man," of which the Black population group – men, women and children – must function as the negation.
>
> (Wynter in Thomas 2006, 25)

This for me raises questions of how a normative and universalist feminist theorizing confronts just what this means for feminist theory. In light of the ascendancy of Black feminist theory, the visibility of Black women as public intellectuals, the centrality of Black feminist theory and scholarship to both Black Studies and Feminist Studies necessitates a repositioning of Black feminisms. I wonder, for example, if intersectionality has not been mobilized to contain the foundational challenge to feminist theory which Black feminist theorizing poses.

Any uncomplicated positioning of Wynter in regard to feminist theorizing risks obscuring Barnes' foundational engagement on Wynter, feminism and gender in which Barnes declares Wynter to be "the reluctant matriarch of Caribbean feminisms" (Barnes 1999) or Wynter's now iconic Afterword to the 1990 pioneering publication, *Out of the Kumbla*. Barnes (1999) locates Wynter's reluctance in the tension between nationalism and feminism. As Reddock (2013) has demonstrated, women's/feminist organizing in the Caribbean often had its genesis in nationalist and left politics, particularly women's dissatisfaction with sexist treatment therein. Carole Boyce Davies' (2018) more recent reflections also offer a more complex relationship between Wynter and feminism. Further, Xhercis Méndez and Yomaira Figueroa (2020) argue that Wynter's "critique of mainstream liberal feminism has provided language to dismiss the concerns articulated and work produced by women of color" (64). Reddock's (2013) assertion that "no other word in modern times has been so vilified for its European origins as feminism" (245) reminds us that those who wish to dismiss the concerns and scholarship of women of color hardly need or needed Wynter to do so. Nonetheless, it is to this complex and complicated relationship with gender, feminism and feminist theory to which I return.

In her conversation with Greg Thomas, Wynter asserts:

> It is not that I am against feminism: I'm appalled at what it became. Originally, there was nothing wrong with my seeing myself as a feminist; I thought it was adding to how we were going to understand this world. If you think about the origins of the modern world, because gender was always there, how did we institute ourselves as humans; why was gender a function of that?
>
> (2006, 23)

She then explicitly cites an example of her work on Black women being deliberately suppressed by other feminist scholars in maintenance of a particular feminist orthodoxy. Andaiye ([2009] 2020) too shares a set of concerns with feminist orthodoxy and praxis which are not unrelated. Citing Reddock, she notes that the Caribbean Association of Feminist Research and Action "was founded to give sex 'primacy' over 'class'" (236). However, she argues that the most promising and *unfulfilled* aspect of its mandate was "to develop an approach to women's problems from the perspective of race, class and sex" rooted in an understanding of how capitalist relations both benefited and produced these unequal relations of power (Andaiye 2020, 236). Andaiye also outlined Caribbean feminist exclusions around race, particularly the exclusion of Indigenous women, class – epitomized in a shift away from prioritizing poverty elimination – age, dis/ability, gender identity and sexuality. This refusal of a feminist party line and willingness to subject feminism to (self)-critique is a defining part of Andaiye's praxis. I find it fruitful to think of Wynter's engagement with normative feminisms through Andaiye's praxis of (self)-critique. Indeed, Andaiye asserted that formal feminist organizing in the region is based on exclusion and therefore utterly unsuited for the task of confronting the multiple, interlocking and co-constituting crises which threatened our very survival. Might part of this failure not also be rooted in another kind of orthodoxy – in the intellectual, political, institutional and bureaucratic containment of feminism and the complicity of feminist scholars and activists therein? Or as Andaiye asserts:

> I think that if I assume that some of us meant what we said in the beginning, which had to do with the interrelationship of oppressions and transformation, then it is very disturbing to see how that translated into all of us becoming either theoreticians of this view or consultants on it.
>
> (Andaiye in Scott 2004, 209)

Andaiye (cited in Scott 2004), in dialogue with David Scott, was quite clear about the limitations of governance feminism, articulating a scathing critique of Caribbean feminisms as centrally concerned with "the professional woman managing and controlling" (210). She indicates that she does not identify as a feminist and outlines why in an analysis which resonates with the concerns raised by Wynter:

> Feminist politics in the Caribbean is a politics which does not rise up as one when the United States destroys Dominica's banana industry. That's not a feminist problem or feminist business. Cuba is not feminist business. Venezuela is not feminist business, even though working-class women of color are leading the process in Venezuela. [...] I don't want any part of a politics where that is not my business.
>
> (Andaiye in Scott 2004, 212)

For both Andaiye and Wynter it is clear that while "gender" is significant, "it is *not* the code itself" (Wynter in Scott 2004, 186).

The uncomplicated positioning of Wynter as a Black American feminist scholar also extracts her from her Caribbean roots in Cuba and Jamaica, despite the clear centrality of the region to her theorizing and the value of reading her alongside other Caribbean feminist scholars. I was recently asked to review an article on Wynter and gender which explicitly stated, "While recognizing the important role that Wynter plays in Caribbean theory, in this essay, I primarily turn to the uptake of Wynter's work in U.S. scholarship." This move is emblematic of the imperialist and colonizing epistemological protocols of US academia as empire. The irony here is that I am being asked to do unpaid intellectual labor for someone in the US academy who devalues my scholarship by virtue of my location outside of the United States.

Elsewhere I have placed Wynter's genre in conversation with Caribbean feminist theorizing of gender and have explored what her theorizing of the human means for masculinity studies (Haynes 2012, 2016). I came to this work out of an interest in Caribbean knowledge on gender and sexuality which circulated among academics, activists and ordinary folks writing in popular spaces such as newspapers and online message boards of the early 2000s. From my vantage point as an adolescent in the 1990s, discussions on male marginalization dominated public debate on gender. Our Sixth Form Spanish teacher brought in editorials on the marginalization of men and boys from national newspapers to be studied alongside *La Casa De Bernarda Alba* and *Yerma* by Federico Garcia Lorca. Wynter's 1968 *The House and Land of Mrs. Alba* was not presented to us at that time. Even though, as Imani Owens (2017) writes, "Wynter's version amplifies the social tensions of Lorca's play by considering the intricacies of race, gender, and class. Whereas Lorca's play concerns the conflict between the Alba daughters, here Mrs. Alba's black servants hold center stage" (52). The Black servants discuss men who "warm their old eyes on young women's breasts," "walk off leave us with children to put food in their mouth" and ask

> [w]hich of the planters not married, ugly, old, bald or fat? Which of their sons don't run away from the land and the mortgage the land carry like a hamper on its back? Which of them don't go away to England these days, get their education and never set a foot back?
>
> (1968, 56)

Asymmetrical and unjust relations of gender are not just evident but inherent in empire, and the race, class and color stratification of the plantation society and economy.

The editorials we examined in class focused on boys in education: we girls were outperforming boys in a space that was rightfully theirs. When you consider that the educational reforms of the 1950s, 1960s and 1970s served to open up education to Black Barbadians of any gender, that by the 1990s the central debate on education would be about the overperformance of Black girls and the underperformance of Black boys is intriguing. I want to suggest here that this is one of the ways in which the coloniality of gender shows up, where race inheres even as the discourse presents itself in gendered terms. Elite girl schools in the 1990s were just as Olive Senior describes them in her 1985 poem Colonial Girls' School: "Borrowed images willed our skins pale/ muffled our laughter/ lowered our voices/ let out our hems/dekinked our hair/denied our sex in gym tunics and bloomers." I felt alienated by the racial and class politics which the school inherited at its founding in 1883 and which, of course, unfolded in very gendered ways. My only claim to belonging would rest on good academic performance. However, girls were routinely sexualized in this debate, viewed as sexually distracting boys and our educational achievements were rendered suspect. It was argued that teaching methods were designed for girls who could thrive in an environment of infinite tedium and rote-learning, whereas boys required intellectual stimulation, discovery and adventure. That the majority of teachers were women was understood to mean that boys were discriminated against, and girls offered an advantage. By the time I entered the PhD program in 2007 my interest in how certain knowledges about gender and sexuality consolidated themselves and were subsequently put to work was a result of living these contradictions.

Outside of the graduate syllabi in our required courses, I found Carole Boyce Davies: *Out of the Kumla* published in 1990 (with Elaine Savory), *Black Women, Writing and Identity: Migrations of the Subject* in 1994 and *Left of Karl Marx: The Political Life of Black Communist Claudia Jones* published in 2008. These texts were all invaluable in seeking to make sense of a Caribbean feminist intellectual tradition which blurred boundaries between theory, activism and creative work. Reading Wynter's Afterword in *Out of the Kumbla*, I was intrigued by the question she raises about the "absence of Caliban's woman." It was rooted in an understanding of systems of knowledge, their role in maintaining the epistemic order and its material effects. Her work was valuable not just for theorizing of gender (or genre) but for theorizing knowledge and power. Rowley's (2010b) critique of Caribbean feminist advocacy as something done on behalf of a group of women – Black, working-class, maternal – by another and Robinson's (2007) critique of the heteronormativity of Caribbean feminist politics seemed illuminated by Wynter's theorizing.

Gender Somethings in and as Context

Greta Olson and Mirjam Horn-Schott (2018) argue that gender as a concept is long past its prime and that Gender Studies itself peaked about 20 years ago. They credit queer theory, sexuality studies, intersectionality and masculinity studies with diversification of the field of Gender Studies and view them as having established fields or sub-fields in their own right, with gender itself coming in for sustained critique from a number of locations. They declare decolonial and materialist feminisms to represent a feminist "now," positing decolonial queer feminisms as the future, a move which feels like yet another progress narrative.

The afterlife of gender looks different depending on your vantage point. Caribbean governments have been responding to the globalized imperative of gender mainstreaming through the establishment of national gender policies since the early 2000s (McFee 2017). Gender specialists continue to work throughout the Global South as part of the implementariat of international development with its attendant internalized hierarchies, invisibilized labor and gender equality imperatives (Warne 2020). Indeed, gender remains the claim to expertise of governance feminism (see Halley et al. 2019). The claims made in its name in the field of Gender and Development and by international NGOs may be quite different from or even inimical to the dominant approaches/trends in the academy.

My adolescent experience of increased public debate on gender is echoed in the concerns of Caribbean feminist scholars and activists. They felt that at the very moment that gender entered the everyday language of policymakers and media pundits, this moment is marked by the pervasiveness of *gender somethings* and the elusiveness of justice (Robinson 2004, 612), "rhetorical practices of gender mainstreaming and the lack of substantive moves toward gender equity," the emergence of new gendered subjectivities and the resilience of old forms of domination (Rowley 2010a, 86). Everybody *talks gender* now. Not only has there been increased knowledge production about gender relations in the academy both by feminist scholars and others across the disciplines; and a burgeoning body of scholarship on Caribbean masculinities which seeks to examine "men as men"; in the popular domain there is also a self-conscious interest in gender evidenced by online fora dedicated to the topic. While Caribbean state managers have avowed their commitments to gender equality in order to ensure good standing within the international community – this gender equality has come at the expense of women's voices (Robinson 2004, 615). It has also come to engender the absurd (Pateman 1986, 4).[1] For example, during the 2008 Emancipation Day celebrations, the Barbadian Minister of State in the Ministry of Culture insisted that henceforth, folklore character,

Mr. Harding would be gender neutral (Nation News, July 30, 2008). Mr. Harding embodies the "hard times" associated with the legacies of enslavement, the physical labor of sugar cane cultivation and harvesting, economic hardship and marginalization in a plantation economy. These hard times are represented as a white male of the planter class – suit, tie, top hat and pipe – and burned in effigy during the Crop Over celebrations. Just what a gender-neutral Mr. Harding was intended to symbolize or achieve remains unclear. How might a gender-neutral Mr. Harding erase either the violence of colonization or the desire to burn Mr. Harding?

As I write, a conservative Christian platform of transphobia, homophobia and misogyny mobilized against "gender" has been fully consolidated in many parts of the Caribbean (see Lazarus (2011, 2012) for Jamaica and Wallace (2017) for The Bahamas). Some of the responses to Barbados' 2023 updated National Grooming Policy with its core value of "gender neutrality," even as school uniforms are gendered, saw some detractors suggest that the changes signal an advancing "LGBT agenda": "*Rasta easy to fool this hair freeup is to push the LGBT agenda just wait till these trans step out to school with weave wanna really think the government doing it for rasta*," to cite but one comment circulating via social media. What is clear, however, is that public discourse on gender has reached abysmal levels of absurdity and incredulity. It is the sheer absurdity of these *gender somethings*, their implications in misogynistic, transphobic and homophobic exclusions, and their pervasiveness both in the academy and wider society that forces an examination of the epistemological and political foundations on which they are produced.

In many ways this consciousness of gender is owed to Caribbean and transnational feminist scholarship and activism which insisted on the usefulness of gender as a tool of analysis for understanding the social world. Tracy Robinson's (2004) *gender somethings* captures the dissatisfaction expressed by other Caribbean feminists at "the ritualistic invocation of gender" (Trotz 2007). These *gender somethings* reveal a broad-based gender consciousness that does not require any commitment to working toward gender equity (Mohammed 2003), and the divorcing of gender from women and feminism (Barriteau 2001). *Gender somethings* points to ubiquity, conceptual promiscuity and nebulous meanings. Robinson (2004) defines *gender somethings* as characterized both by their exclusion of questions of transgressive sexuality or sexual rights and freedom and their signaling an understanding that whereas this "thing" called gender equality is important, especially for a country's international reputation as a modern nation, other things are more important. As knowledge-practices, *gender somethings* are produced by power and are productive in their own right, especially of new ontological positions.

If Robinson's nomenclature is most apt, Barriteau (2001, 71) provides the most sustained critique of the way in which gender has been mobilized by the twentieth-century Caribbean state to ensure that "[i]deologically, relations of gender are at their worst for Caribbean women." Barriteau's analysis of how the distorted work of feminist scholars from outside the Global South became codified in development theory and practice, and resulted in an avalanche of gender "manuals, models and approaches," shows how power/knowledge becomes operationalized and actualized in people's everyday lives (Barriteau 2001, 85). It also foreshadows the lack of institutional clarity with which women's desk/bureau transitioned to gender bureau and ultimately the co-optation of feminist work by the state. This she refers to as "the epistemological hijacking of the century" (Barriteau 2001, 90). Gender as a tool of analysis was replaced by its usage as a less politicized term, whose meaning though nebulous was understood to be antithetical to a feminist politics. It was also simultaneously understood to refer to male/female relations (Barriteau 2001, 87).

Barriteau (2004) argues that the shift from "woman" to "gender" in popular discourse, the academy and within the women's machineries (now gender bureaus) has not been adequately problematized and we are therefore unaware of just what has been lost in the shift. Through a genealogy of the word gender, she reminds us that despite the fact that gender is now a depoliticized, sanitized term, popularly understood as "women and men," among several other competing definitions, it is indeed a feminist conceptual tool. She demonstrates how the term has been stripped of its feminist roots and the ways in which the shift has been represented as a shift from "women" to "men." To introduce Wynter's theorizing into this debate is to ask questions of Caribbean feminist attachments to gender and to the optimistic epistemology which inheres in moves to fix and delineate its meaning.

Teaching Feminist Theories of Gender in the Caribbean

At the Cave Hill Campus, I teach a student body which is majority Black Barbadian and Eastern Caribbean women who are part-time students from working-class or lower middle-class backgrounds. It is quite common for students to be the first in their families to attend university. The campus has been forced to confront sexual harassment, misogynistic, homophobic and transphobic violence and to increase access to students with disabilities. Concerns about the need to increase young men's enrollment in tertiary education have also been consistently articulated.

It is hard work convincing students in a Gender and Development Studies program that gender is not an epistemological given, and that feminist conceptualizations of gender are neither static, singular, uncontested nor uncompromised, especially when they have grown up in a media-saturated era of *gender somethings*. I center the work of Black and Caribbean feminist scholars such as Hortense Spillers, Cecilia Green, Xhercis Méndez, Michelle Rowley, Eudine Barriteau, Rhoda Reddock and Patricia Mohammed in dialogue with Sylvia Wynter in my teaching. In teaching Spiller's foundational "Mama's Baby, Papa's Maybe" to an undergraduate class last semester, I presented the following excerpts accompanied by an image of the hold of a slave ship:

> Let it now be supposed ... further, that every man slave is to be allowed six feet by one foot four inches for room, every woman five feet ten by one foot four, every boy five feet by one foot two, and every girl four feet six by one foot. "... five females be reckoned as four males, and three boys or girls as equal to two grow persons."
>
> (cited in Spillers 1987, 72)

> Under these conditions, one is neither female, nor male as both subjects are taken into "account" as *quantities*. The female in "Middle Passage" as the apparently smaller physical mass, occupies "less room" in a directly translatable money economy. But she is, nevertheless, quantifiable by the same rules of accounting as her male counterpart.
>
> (Spillers 1987, 72)

A student's hand shoots in the air. The excited expression on her face is one of visceral understanding of the arguments made by Spillers and Wynter regarding race and gender in the New World. I pause to give her the floor. Students, of course, make the best teachers. "What if the woman needs more room? They are taking all women as the same and they are giving them less room than the men." I am stunned into a combination of silence, shame and disappointment. I am forced to recognize that students come into Gender Studies classes with an expectation that the class will be about gender, that is, women's position relative to men's. In this frame, the dehumanizing nature of slavery can go unremarked, in favor of an analysis of the differential treatment of enslaved men and women. We have indeed not moved *beyond* gender after all. At the secondary schools from which they graduated, history has been removed as a compulsory subject, and most come with little knowledge of Caribbean history. Gender, however, is a frequent topic of debate on mainstream and social media. Students come with a particular knowledge of the key concerns that are ventilated in public and with popular feminist or post-feminist knowledge which circulates online.

To correct these gaps, I begin with Cecilia Green's fine sociological history of gender and labor in the Caribbean. In particular, I teach her essays "'The Abandoned Lower Class of Females': Class, Gender and Penal Discipline in Barbados, 1875–1929" (2011) and "Disciplining Boys: Labor, Gender, Generation and the Penal System in Barbados, 1880–1930" (2010). Green's work provides the empirical evidence that the simultaneous exclusion of Black women from normative gender categories and their sexing as female creates a lose-lose situation for Black women. Their exclusion from the normative category of woman makes them available as hyper-exploited laborers and "female flesh ungendered" to use Spillers' terms. At the same time, during the mass migration of Barbadian men to Panama, the colonial authorities view them and Black boys as abandoned dependents. In this assumed absence of patriarchal control, colonial authorities target women for incarceration and boys for violent, physical discipline. Black women and boys emerge as a problem. Students make connections with how race shapes gender. They recognize that this often shows up currently as respectability politics. This work demonstrates the co-constitutive nature of race and gender in shaping negative outcomes for working-class boys. They recognize that being "a good woman of your kind," to paraphrase Wynter, is inherently a losing position. They assert the need for gender self-determination. They find in the collective work of scholars such as Wynter, Spillers and Green the grounds on which to "divest from heterosexual precepts" and to articulate solidarities with queer and trans folks or to claim these identities for themselves in the classroom (Robinson 2007, 127). This banishing of Blackness from the human through norms of gender and sexuality opens up for them productive and liberatory possibilities and grounds for solidarity across gender identity, dis/ability, sexuality and class.

Students also appreciate the introduction of recent material drawn from the media and everyday life which compliments what we learn from Wynter's and Green's historical analysis. For example, as we interrogate the category "woman" I frequently share Mark Critchley's 2016 article published online in the UK edition of The Telegraph with its headline and sub-title: "Rio 2016: Fifth-placed Joanna Jozwik 'feels like silver medallist' after 800m defeat to Caster Semenya: Jozwik controversially claimed that she was proud to have finished as the 'first European' and 'second white'." Students analyze the significance of being "the first European" and the "second white" in the world's most prestigious athletics competition where neither Europeanness nor whiteness are ostensibly competition criteria. They make connections with the theorizing of Wynter and Spillers which help them understand Jozwik's disqualification of South Africa's Caster Semenya, Burundi's Francine Niyonsaba and Kenya's Margaret Wambui as *women* and as Olympic medalists. They note that Jozwik's claim that one need only *look* at the three top-placed athletes helps them to understand the visual racist logics of gender. As one student wrote, "Up until I started taking this class, I had trouble explaining my experiences." Presenting the scholarly and theoretical material in relation to the everyday events encourages a reading practice that seeks to both understand the work on the page and its possibilities off of it.

I frequently ask students to prepare analytical response papers to the assigned readings and to present their critical engagement in an oral presentation supported by visual material. They also prepare discussion questions so that they lead and facilitate the class discussion. However, I've never used this strategy with Wynter's work and tend to both present the work and lead the discussion myself. Perhaps I've internalized the view that Wynter is "difficult." In teaching Sylvia Wynter, I draw on Méndez's (2015) "Notes Toward a Decolonial Feminist Methodology" and Patricia Fox's (2006) "After Man, Womens" which asks if Black woman is an oxymoron. One student is silent for the entire lesson. She is busy making notes. At the end of the class, she reads verbatim from the notes she has written. She has in fact been writing a rebuttal. She has understood me to be disparaging Black women and denying our womanhood and humanity. For her, the negation of black humanity within white imagination equals the negation of blackness and she will have none of it! I wonder about this sense of loss felt on recognizing that normative gender categories depend on whiteness. Should it not instead provoke a feeling of liberatory possibilities given our understanding of these categories as oppressive? For as Xhercis Méndez (2015) outlines, there is a lot that has to be given up in order for Black women to argue ourselves into normative gender categories. In any case, I am guided by Carole Boyce Davies (2018) who reminds us that: "As with most Wynter essays, it demanded multiple readings, always with the accompanying feeling when first encountering a Wynter essay, that full meaning was just beyond one's grasp" (838). The student is concerned that Wynter's arguments invalidate her research project which focuses on Black Caribbean women and beauty.

I have watched our graduate students have to navigate critiques regarding their decision to focus on Black women's lives in their research. During a seminar, examination one of the assessors told a student who was researching mothering that Gender Studies demanded that her research examine the lives of *both* mothers and their presumably male heterosexual, cis-gender partners. In another instance, a student was forced to account for both an absence of Caribbean (cis-gender) men's experiences in her research on young Afro-Guyanese women's sexualities, as well as Indo- and Indigenous Caribbean women's. Three kinds of trends converge here: the claim that Blackness is particular, that deployments of gender must of necessity include both women and men, as well as concerns about racial/ethnic exclusions such as those articulated by Andil Gosine (2009, 5) who criticizes the

> commitment of even very critical and interrogative scholars to ethnic and racial compartments, at the expense of recognizing the dynamic present and history of the Caribbean. A lot of the emerging studies of Caribbean sexuality, for instance, leave completely unconsidered the experience of Indo-Caribbean people and indentureship.

Students themselves also push back against what they see as too much discussion about "race" in a Gender and Development Studies classroom. I've had students assert that Black feminist scholarship narrates a particularized Black experience with little applicability outside of Black women's lives. They want theories which travel. They have invoked the multi-ethnic and multi-racial composition of their countries of origin as the reason why they find Black feminist theorizing in general and Wynter in particular to be of very little use. One student in particular noted that Butler's performativity of gender has broad applicability in the multi-racial context of her home country but that Caribbean feminist theorizing of gender as advanced by Barriteau and Mohammed, and Wynter's genre do not. That students may initially reject what is Black or Caribbean as too specific, too particular, too parochial, is, of course, no surprise. On the other hand, in student evaluation of courses the emphasis on Caribbean scholarship and context is often cited as what they found most valuable about the course. Another student explained that during her work with one of the region's gender machineries, she was explicitly asked not to include anything about feminism or feminist theory in a workshop on gender which she was asked to deliver. On graduation students report *both* a demand for the expertise they bring from the field of Gender Studies and a request that such knowledge be presented without mentioning feminism or the existence and human rights of lesbian, gay, transgender, bisexual, queer and intersex Caribbean people.

As Imani Owen (2017, 50) suggests, "the significance of Wynter's bilingualism on her scholarly and creative sensibility is not adequately appreciated." Katherine McKittrick (2021) calls to attention to Wynter's engagement with science in creative and theoretical work, this blurring of boundaries also demands creativity in the classroom. Wynter's own creative life as a dancer, actor, playwright and novelist as well as her theorizing of popular and folk culture demands multiple fluencies in the teaching of Wynter in Gender and Development Studies. On the syllabus I've never included her plays such as *The House and Land of Mrs. Alba* or *Maskarade* even as they are also revealing of Wynter's thoughts on gender. It is my intention to remedy this.

Wynter and Feminist Critiques of Gender

I am guided by bell hooks' (2000, xi) caution that "Often when individuals talk or write about contemporary feminist movement they make it seem as though there was a set body of feminist principles and beliefs that served as a foundation from the very beginning." In teaching feminist theorizing of gender, I therefore seek to place multiple genealogies of gender in riotous conversation with each other, recognizing that for many feminist theorists, the construction of gender "as a separate social/cultural/psychological category" not only allowed them to examine and highlight the specificity of sexist oppression but just as importantly, to claim exclusive epistemological terrain (Lykke 2010, 107). Feminist theories are "intellectual tools by which historical agents can examine the injustices they confront, and can build arguments to support their particular demands for change" (McCann and Kim 2003, 1). If we agree with Wynter (2006), however, that "to be 'modern,' to be 'academics,' we are all Westerners," i.e., that we are implicated in maintaining the current order of knowledge at the root of the apocalyptic climate crisis, global anti-Blackness, Indigenous dispossession, the neo-liberal economic order and pervasive gender-based violence, then not only must feminist scholars and activists abandon claims to innocence but we must also recognize how feminist theory supports rather than subverts the current order of knowledge. Interrogating the stories which feminists tell about gender is a helpful place to start (Hemmings 2011). This genealogy also demonstrates that Wynter is not unique in critiquing gender and asks what might be gained from situating her critique in dialogue with feminist self-critique. Marquis Bey's (2016) critique that Black feminist thought, "the framework and field have yet to fully embrace black trans and gender-queer women, which reflects a serious infraction of its commitment to liberation" (37), may be said of feminist thought more broadly and is clearly evident in the multiple feminist theorizings of gender which I examine here in which only cis-gender women and men are imagined in a frequently ahistorical and self-evident sense.

I am also motivated by Keguro Macharia's (2020) insistence that "one of the most tedious and necessary things is to figure out is the exigency for conceptual work." Macharia proposes that we ask the following questions of our concepts:

- Why was a particular concept or method developed?
- What was it engaging?
- What work was it supposed to do?
- What work did it do?
- Why is it still necessary?
- Why might it be exhausted?
- Which people did it assemble?
- What happened when they assembled?
- If the concept or method is lost, what caused that loss?
- What lessons can be taken from the history of that concept or method and who it assembled and how it assembled?

I insist that students ask these questions of any concept or method which they take up. Of necessity this requires that they move between debates within the academy as well as the social world which they are seeking to investigate. It allows them the understand the particular histories, geographies, analytical value and limitations of the concepts they confront in the classroom. It also allows them to make wise judgments of the universalist, racist and transphobic baggage which may cling to some concepts long after the academy may have declared itself to have moved beyond such origins. I also wish to unsettle students' confidence that they *know* what gender is or that they can know a priori just what it is at stake when it is invoked.

Feminist conceptualization of gender as a significant variable in human social organization has been heralded as "one of the most innovative insights of the last decades" (Wieringa 2002, 17). The most oft-repeated feminist story of gender is that prior to the 1970s the term gender was not used to refer to (relations between) people but was used in linguistics to categorize masculine and feminine nouns (McCann and Kim 2003). A corrective to this narrative is that second-wave feminist theorizing of gender came out of feminist writings in the 1940s, a period generally considered one of feminist inactivity in terms of a lack of feminist activism (Delphy 2003, Tarrant 2006). This understanding, however, discounts the activism of African American women activists in the 1940s and the relationship of questions of gender and sexuality to what came to be known as the Civil Rights movement (see McGuire 2010). Carole Boyce Davies' (2008) fine biography of Claudia Jones also challenges these claims of Black feminist inactivity during this period as does Erik McDuffie's (2011) brilliant and foundational study of Black communist women. Other feminist genealogies of gender credit the sexology of the 1950s with its medicalized violence against intersex people for feminism's sex/gender distinction (Germon 2009, Meyerowit 2008).

Gerda Lerner (1986, 238) opines that the traction gained by "gender" in both feminist theory and the media was due to its disassociation from the "nastiness" of sex. Gender then came to demarcate a certain distance from questions of sex and sexuality in contrast to the ways in which it had previously been deployed. Braidotti (1994) also argues that as a "scientific-sounding term," gender allows feminists to claim space within the academy but at the expense of their political agenda (151). She argues that the success of Gender Studies in the academy is a result of the "reassuring" and apolitical overtones of gender.

On the establishment of the Centre for Gender and Development Studies at the University of the West Indies, the politics of naming had to be negotiated: should the focus be on Women's Studies, Gender Studies or Gender and Development Studies? Some rejected gender on the grounds that it was "more neutral, non-controversial, and socially acceptable" and therefore inadequate to the liberatory task at hand (Massiah et al. 2016, 114). The prevailing voices argued that gender was in fact, "a complex and comprehensive concept" (Massiah et al. 2016, 114).

Since the 1980s, Joan Scott has highlighted the lack of conceptual clarity around the term gender among feminists themselves. Gender has been used as a synonym for women, as a neutral term which distances itself from the so-called stridency of feminism, and as a more "scientific" and academically acceptable term akin to class (Scott 1988). It has also been used by feminist theorists "as a role, as a social category, as a social practice, as performance, as social structure, and sometimes a combination of these" (Foster 1999, 433).

Despite gender's conceptual plurality, gender represents the feminist claim to exclusive epistemological terrain and feminist naming of the specificity of a generic "women's subordination":

> The term "gender" is part of the attempt by contemporary feminists to stake claim to a certain definitional ground, to insist on the inadequacy of existing bodies of theory for explaining persistent inequalities between women and men.
>
> (Scott 1988, 44)

Biologist Donna Haraway (1991) argued that feminist theorizing of gender maintained Western binarism with its "appropriationist logic of domination" (198) or the understanding that gendered relations of power are not natural and are therefore under our control. Feminist philosopher Rosi Braidotti (1994) argued that gender represented a shift in focus away from women which ran counter to a feminist political and epistemological praxis. She proposed "sexual difference theories" as opposed to theories of gender. Sexual difference theories were themselves criticized for suggesting that sexuality should be the primary focus in addressing and combating women's subordination (Foster 1999). Criticisms of the liberal focus on gender are echoed in Wynter's (2018) insistence that feminism requires an autonomous ground beyond liberalism and Marxism. Wynter's analysis of Man goes beyond the contributions of early feminist theorizing of gender to reveal the economistic and race categories which inhere in its conceptualization. Nivedita Menon's (2009) work on India which tracks gender from the 1970s has revealed that in the 1990s feminist deployments of gender were productively fractured by discourses of sexuality and caste while state discourses of gender served to domesticate the term.

Feminist genealogies of gender must confront gender's multiple meanings and lack of conceptual clarity within feminism. They must also account for its Westerncentrism (Oyěwùmí 1998), "appropriationist logic of domination (Haraway 1991, 198)," complicity with intersex erasure and desire for epistemological terrain and institutional power within academia. We cannot assume that these histories have been excised despite narratives of progress and waves in feminist theorizing. Wynter challenges both the hegemony of gender within feminist theorizing and how gender is defined, questioning the foundational assertion that "woman" is Man's other. She refutes that gender is epistemologically primary, nor is it the primary way of signifying relations of power. The uptake of intersectionality can perhaps also be attributed to the fact that it displaces the hegemony and primacy of gender in feminist theory without denying its salience. The promise of gender as a depoliticized, respectable area of analysis where developing countries are placed "under Western eyes" has been fully realized. Transnational feminist critiques such as those advanced by Jacqui Alexander (2005), Michelle Rowley (2010c, 78) and Nicole Charles (2022) which "do not give the category "gender" a priori primacy" but rather, as Rowley (2010c, 78) insists "the salient analytical markers must, above all else, be determined by the fields of play in which power reveals itself" serve to displace gender's primacy while maintaining its salience. A move which I do not read as at all at odds with Wynter's theorizing of genre. In *Cistem Failure*, Marquis Bey (2022b) invites us to take as our point of departure the accumulated knowledge about gender offered by (Black) feminist scholars and queer theorists:

> We cite the radical feminist knowledge that *sex's difference from gender makes possible the account of un/alignment that constitutes cisgender and transgender as discrete and self-enclosed identities*. We cite that *whiteness is constitutive of binary gender as a construct*. We cite how we wish to be *against and beyond the constitutively white settler binary cis gender symbolic and social order*. We cite that *this problematizing of gender places her, the black woman, out of the traditional symbolics of female gender, and it is our task to make a place for this different social subject*. And we cite: *There is no body, no sexuality and, simply put, no sex outside the long history of Western imperialism's shattering of the world.* All right, then, let's begin from here.
>
> (2022, xv, emphasis in original)

This is a politics of possibility which conceives of gender in systemic terms and seeks a radical *beyond.* If we take seriously Baldwin's (2022) exposure of the colonial and plantation logics of the contemporary neo-liberal university and the racist and gendered effects of its diversity and inclusion mandates, then of necessity we must regard with suspicion the legitimizing work that "gender" as a concept was called on to do. In teaching Gender and Development Studies, I always make sure that students are aware of these multiple, competing, contested and compromised deployments of gender.

Against Resolution

I have resisted suggesting that replacing gender with Wynter's genre would resolve the ethical, political and conceptual stakes involved in feminist theorizing of gender. This would be a naïve view. In addition, as some make the claim that feminist scholarship has moved *beyond* gender for at least the last 20 years, such a move would be redundant. If we are indeed beyond gender, then the matter is settled. I do not read Wynter as dismissing the value of Black and Caribbean feminist scholarship, and I am therefore neither articulating a defense of Wynter nor of Caribbean scholars. I have sought instead to highlight student experiences both within the classroom and in their work outside the classroom in the field of Gender and Development, where they encounter preconceived understandings of just what gender is and collective agreement on its salience and primacy. My hope is that the heterogenous and ethically and politically compromised theorizing of gender which I have presented would at least serve to undo gender's coherence as a feminist concept and displace its *a priori* primacy. I have also sought to provide examples from transnational and Black feminist praxis which do just this. This goes beyond Sandra Harding's (1992) assertion that the analytical categories of feminist theory are inherently unstable to consider how they are always already compromised. All of our concepts are! It means our deployments of them must be contingent, specific, free of naïve claims to innocence, and open to acknowledgment of their analytical limits. It also requires that the intellectual and political history of the concept be laid bare. Perhaps, part of the way we trouble the governing system of meaning is to seek to relinquish the arrogant expectation that we could ever single-handedly apprehend the social world in a single concept or theory. In their critique of Wynter, Méndez and Figueroa (2020) defend the value of concepts such as patriarchy and gender, parting ways with Wynter in the moments they read her as dismissive of feminism, patriarchy and gender. They affirm that the experiences of women of color necessitate both feminist politics and the conceptual tools of feminist theorizing. They highlight the existence of patriarchal relations of gender outside of colonization. Perhaps Women's/Gender/Feminist Studies after Wynter means moving away from individualizing and isolating her and instead locating her within broader Black, Caribbean, feminist and anti-colonial intellectual, political and artistic traditions. This has been my approach to teaching Gender and Development

Studies in the classroom. There is no need to have the last word on Wynter, gender and feminism nor for Wynter to have the last word. The tensions are generative. More significantly, arresting the homophobic, transphobic and misogynist backlash against the perceived ascendancy of "gender" cannot be achieved by authoritatively fixing gender's meaning, returning gender to its feminists roots – plural, contested, compromised – or declaring ourselves to have moved beyond gender. But rather, in the ongoing, often unrecognized work of seeking to harness the emancipatory possibilities of epistemic ruptures of the governing system of meaning to use Wynter's terms as well as ruptures in gender systems, to use Barriteau's terms – the productive possibilities of *cistem failure* (Bey 2022).

Notes

1 Feminists have highlighted the ways in which gender-neutral policies often work against the interests of women or engender absurdities. Carole Pateman, "Introduction: The Theoretical Subversiveness of Feminism," in *Feminist Challenges: Social and Political Theory*, eds. Carole Pateman and Elizabeth Gross (Sydney, London and Boston: Allen & Unwin, 1986), 8.

References

Alexander, M. Jacqui. 2005. *Pedagogies of Crossing: Meditations on Feminism, Sexual Politics, Memory and the Sacred.* Durham: Duke University Press.

Andaiye. 2020. *The Point Is to Change the World: Selected Writings of Andaiye*. London: Pluto Press.

Baldwin, Andrea N. 2022. *A Decolonial Black Feminist Theory of Reading and Shade: Feeling the University*. Abingdon Oxon: Routledge. https://www.taylorfrancis.com/books/9781003019442.

Barnes, Natasha. 1999. "Reluctant Matriarch: Sylvia Wynter and the Problematics of Caribbean Feminism." *Small Axe*, 5: 34–47.

Barriteau, Eudine. 2001. *The Political Economy of Gender in the Twentieth Century Caribbean*. New York: Palgrave.

Barriteau, Eudine. 2002. "Gender? What It Is? What Is It Not? A Genealogy of the Concept of Gender and Its Relevance for Policy Makers." Cave Hill: University of the West Indies.

Barriteau, Eudine. 2004. "Theorizing the Shift from Women to Gender in Caribbean Feminist Discourse: The Power Relations of Creating Knowledge." In *Confronting Power, Theorizing Gender: Interdisciplinary Perspectives in the Caribbean*, ed. Eudine Barriteau, 27–45. Kingston: UWI Press.

Bey, Marquis. 2022a. *Black Trans Feminism*. Durham: Duke University Press. https://www.jstor.org/stable/10.2307/j.ctv23g7wh9.

Bey, Marquis. 2022b. *Cistem Failure: Essays on Blackness and Cisgender*. Duke University Press. https://doi.org/10.2307/j.ctv2rr3g52.

Bey, Marquis. 2016. "The Shape of Angels' Teeth Toward a Blacktrans feminist Thought through the Mattering of Black(Trans)Lives." *Departures in Critical Qualitative Research*, 5(3): 33–54.

Boyce-Davies, Carole, and Elaine Savory. 1990. *Out of the Kumbla: Caribbean Women and Literature*. Trenton N.J: Africa World Press.

Braidotti, Rosi. 1994. *Nomadic Subjects: Embodiment and Sexual Difference in Contemporary Feminist Theory*. New York: Columbia University Press.

Butler, Judith. 1990. *Gender Trouble: Feminism and the Subversion of Identity* New York: Routledge.

Charles, Nicole. 2022. *Suspicion: Vaccines, Hesitancy, and the Affective Politics of Protection in Barbados*. Duke University Press. https://doi.org/10.2307/j.ctv228vqq6.

Davies, Carole Boyce. 2003. *Black Women Writing and Identity: Migrations of the Subject*. London: Routledge.

Davies, Carole Boyce. 2008. *Left of Karl Marx: The Political Life of Black Communist Claudia Jones*. Durham: Duke University Press. muse.jhu.edu/book/69121.

Davies, Carole Boyce. 2018. "Occupying the Terrain: Reengaging "Beyond Miranda's Meanings: Un/Silencing the 'Demonic Ground' of Caliban's Woman"." *American Quarterly*, 70(4): 837–45.

Delphy, Christine. 2003. "Rethinking Sex and Gender." In *Feminist Theory Reader: Local and Global Perspectives*, eds. Carole R. McCann and Seung-Kyung Kim, 57–67. New York: Routledge.

Foster, Johanna. 1999. "An Invitation to Dialogue: Clarifying the Position of Feminist Gender Theory in Relation to Sexual Difference Theory." *Gender and Society*, 13(4): 431–56.

Fox, Patricia. 2006. "After Man, Womens: The Third Sex or Ain't i a Eunuch." In *After Man Towards the Human: Critical Essays on Sylvia Wynter*, ed. Bogues Anthony, 100–30. Kingston and London: Ian Randle.

Germon, Jennifer. 2009. *Gender: A Genealogy of an Idea*. New York: Palgrave Macmillan.

Gosine, Andil. 2009. "Politics & Passion: A Conversation with Gloria Wekker." *Caribbean Review of Gender Studies*, 2009(3): 1–11.

Green, Cecilia A. 2011. "'The Abandoned Lower Class of Females': Class, Gender, and Penal Discipline in Barbados, 1875–1929." *Comparative Studies in Society and History*, 53(1): 144–79. http://www.jstor.org/stable/41241736.

Green, Cecilia. 2010. "Disciplining Boys: Labor, Gender, Generation, and the Penal System in Barbados, 1880–1930." *The Journal of the History of Childhood and Youth*, 3(3): 366–90.

Halley, Janet, Prabha Kotiswaran, Rachel Rebouché, and Hila Shamir. 2019. "Preface." In *Governance Feminism: Notes from the Field*, eds. Janet Halley, Prabha Kotiswaran, Rachel Rebouché, and Hila Shamir, ix–xxxviii. University of Minnesota Press. https://doi.org/10.5749/j.ctvdjrpfs.3.

Haraway, Donna. 1991. *Simians, Cyborgs, and Women: The Reinvention of Nature*. New York: Routledge.

Harding, Sandra. 1992. "The Instability of Analytical Categories of Feminist Theory." In *Knowing Women: Feminism and Knowledge*, eds. Helen Crowley and Sue Hinnelweit, 338–54. London: Polity Press in association with the Open University.

Haynes, Tonya. 2012. "The Divine and the Demonic: Sylvia Wynter and Caribbean Feminist Thought Revisited," In Barriteau Eudine and University of the West Indies. *Love and Power: Caribbean Discourses on Gender*. Kingston Jamaica: University of the West Indies Press: Cave Hill Barbados: Institute for Gender and Development Studies Nita Barrow Unit.

Haynes, Tonya. 2016. Sylvia Wynter's Theory of the Human and the Crisis School of Caribbean Heteromasculinity Studies. *Small Axe*, 20(1): 92–112. https://www.muse.jhu.edu/article/610589.

Hemmings, Clare. 2011. *Why Stories Matter: The Political Grammar of Feminist Theory*. Duke University Press. https://doi.org/10.2307/j.ctv1220mp6.

hooks, bell. 2000. *Feminist Theory from Margin to Center*. Cambridge: South End Press,

Jackson, Zakiyyah Iman. 2020. *Becoming Human: Matter and Meaning in an Antiblack World*. Vol. 53. NYU Press. http://www.jstor.org/stable/j.ctv1n6ptnn.

King, Tiffany Lethabo. 2019. *The Black Shoals: Offshore Formations of Black and Native Studies*. Durham North Carolina: Duke University Press.

Lazarus, Latoya. 2011. "Heteronationalism, Human Rights, And The Nation-State: Positioning Sexuality In The Jamaican Constitutional Reform Process." *Canadian Journal of Latin American and Caribbean Studies / Revue Canadienne Des Études Latino-Américaines et Caraïbes*, 36(71): 71–108. https://www.jstor.org/stable/26505390.

Lazarus, Latoya. 2012. "This Is a Christian Nation: Gender and Sexuality in Processes of Constitutional and Legal Reform in Jamaica." *Social and Economic Studies*, 61(3): 117–43. http://www.jstor.org/stable/41803770.

Lerner, Gerda. 1986. *The Creation of Patriarchy*. New York and Oxford: Oxford University Press.

Lykke, Nina. 2010. *Feminist Studies: A Guide to Intersectional Theory, Methodology and Writing*. New York: Routledge.

Macharia, Keguro. 2020. On Twitter. September 7.

Massiah, Joycelin, Elsa Leo-Rhynie, and Barbara Bailey. 2016. *The UWI Gender Journey: Recollections and Reflections*. Jamaica Barbados: University of the West Indies Press and The Institute for Gender and Development Studies, The University of the West Indies.

McCann, Carole R., and Seung-Kyung Kim. 2003. "Definitions and Movements: Introduction." In *Feminist Theory Reader: Local and Global Perspectives*, eds. Carole R. McCann and Seung-Kyung Kim, 12–23. New York: Routledge.

McDuffie, Erik S. 2011. *Sojourning for Freedom: Black Women, American Communism and the Making of Black Left Feminism*. Durham NC: Duke University Press.

McFee, Deborah. 2017. "Narratives, the State and National Gender Policies in the Anglophone Caribbean: Dominica and Trinidad and Tobago." In *Negotiating Gender Policy and Politics in the Caribbean: Feminist Strategies Masculinist Resistance and Transformational Possibilities*, eds. Hosein Gabrielle and Jane Parpart, 109–29. London: Rowman & Littlefield International.

McGuire, Danielle L. 2010. *At the Dark End of the Street: Black Women, Rape, and Resistance- a New History of the Civil Rights Movement from Rosa Parks to the Rise of Black Power*. New York: Alfred A. Knopf.

McKittrick, Katherine. 2006. *Demonic Grounds: Black Women and the Cartographies of Struggle*. Minneapolis: University of Minnesota Press.

McKittrick, Katherine. 2021. *Dear Science and Other Stories*. Durham: Duke University Press. https://doi.org/10.1515/9781478012573.

Méndez, Xhercis. 2015. "Notes Toward a Decolonial Feminist Methodology: The Race/Gender Matrix Revisited." *Trans-Scripts*, 5: 41–59.

Méndez, Xhercis, and Yomaira C. Figueroa. 2020. "Not Your Papa's Wynter: Women of Color Contributions toward Decolonial Futures." In *Beyond the Doctrine of Man: Decolonial Visions of the Human*, eds. Joseph Drexler-Dreis and Kristien Justaert, 1st ed., 60–88. Fordham University Press. https://doi.org/10.2307/j.ctvsf1q4v.6.

Menon, Nivedita. 2009."Sexuality, Caste, Govermentality: Contests over 'Gender' in India." *Feminist Review*, 91: 94–112.

Meyerowit, Joanne. 2008. "A History of 'Gender'." *The American Historical Review*, 113(5): 1346–56.
Mohammed, Patricia. 2003. "Like Sugar in Coffee: Third Wave Feminism and the Caribbean." *Social and Economic Studies*, 52(3): 5–30.
Mohammed, Patricia. 2002. *Gender Negotiations among Indians in Trinidad, 1917–1947*. New York and Hampshire: Palgrave.
Olson, Greta, and Mirjam Horn-Schott. 2018. Introduction: "Beyond Gender: Towards a Decolonized Queer Feminist Future". In *Beyond Gender: Futures of Feminist and Sexuality Studies – An Advanced Introduction*, eds. Greta Olson, Daniel Hartley, Mirjam Horn-Schott, and Leonie Schmidt. London: Routledge.
Owens, Imani. 2017. Toward a "Truly Indigenous Theatre": Sylvia Wynter Adapts Federico García Lorca. *Cambridge Journal of Postcolonial Literary Inquiry*, 4(1): 49–67. doi:10.1017/pli.2016.34
Oyěwùmí, Oyèrónkẹ́. 1998. "De-Confounding Gender: Feminist Theorizing and Western Culture, a Comment on Hawkesworth's 'Confounding Gender'." *Signs: Journal of Women in Culture and Society*, 23(4): 1049–62.
Pateman, Carole. 1986. "Introduction: The Theoretical Subversiveness of Feminism." In *Feminist Challenges: Social and Political Theory*, eds. Carole Pateman and Elizabeth Gross, 1–10. Sydney: Allen & Unwin.
Reddock, Rhoda, 2013. "Feminism, Nationalism and the Early Women's Movement in the English-speaking Caribbean (with Special Reference to Jamaica and Trinidad and Tobago)." In *Caribbean Political Thought: The Colonial State to Caribbean Internationalisms*, ed. Aaron Kamugisha, 245–260. Kingston: Ian Randle.
Robinson, Tracy. 2007. "A Loving Freedom: A Caribbean Feminist Ethic." *Small Axe*, 24: 118–129.
Robinson, Tracy. 2011. "Our Imagined Lives." In *Sex and the Citizen: Interrogating the Caribbean*, eds. Faith Smith, 201–13. Charlottesville: University of Virginia Press.
Robinson, Tracy. 2004. "Gender, Feminism and Constitutional Reform in the Caribbean." In *Gender in the 21st Century: Caribbean Perspectives, Visions and Possibilities*, eds. Barbara Bailey and Elsa Leo-Rhynie, 592–625. Kingston: Ian Randle Publishers.
Rowley, Michelle. 2010a. *Feminist Advocacy and Gender Equity in the Anglophone Caribbean: Envisioning a Politics of Coalition*. New York: Routledge.
Rowley, Michelle. 2010b. "Whose Time Is It? Gender and Humanism in Contemporary Feminist Advocacy." *Small Axe: A Caribbean Journal of Criticism*, 14(1): 1–15.
Rowley, Michelle. 2010c. "Where the Streets Have No Name: Getting Development out of the (Red)?" In *Gender and Global Restructuring: Sightings, Sites and Resistances*, eds. Marianne H. Marchand and Anne Sisson Runyan, 78–98. New York: Routledge.
Scott, David. 2000. "The Re-Enactment of Humanism: Interview with Sylvia Wynter." *Small Axe: A Caribbean Journal of Criticism*, 8: 119–207.
Scott, David. 2008. "Counting Women's Caring Work: An Interview with Andaiye." *Small Axe*, 8(1): 123–217. doi: https://doi.org/10.1215/-8-1-123
Scott, Joan Wallach. 1988. *Gender and the Politics of History*. New York, Chichester, West Sussex: Columbia University Press.
Snorton, C. Riley. 2017. *Black on Both Sides: A Racial History of Trans Identity*. Minneapolis MN: University of Minnesota Press.
Spillers, Hortense J. 1987. "Mama's Baby, Papa's Maybe: An American Grammar Book." *Diacritics*, 17(2): 65–81. https://doi.org/10.2307/464747.

Stallings, LaMonda. 2017. *Funk the Erotic: Transaesthetics and Black Sexual Cultures*. Urbana: University of Illinois Press. https://doi.org/10.5406/illinois/9780252039591.001.0001.

Tarrant, Shira. 2006. *When Sex Became Gender*. New York: Routledge.

Thomas, Greg. 2006. "Proud Flesh Inter/Views: Sylvia Wynter." *PROUDFLESH: A New Afrikan Journal of Culture, Politics & Consciousness* 4. http://www.africaresource.com/proudflesh/issue4/wynter.html.

Thomas, Greg. 2007. *The Sexual Demon of Colonial Power: Pan-African Embodiment of Erotic Schemes of Empire*. Bloomington: Indiana University Press.

Tinsley, Omise'eke Natasha. 2008. "Black Atlantic, Queer Atlantic: Queer Imaginings of the Middle Passage." *GLQ: A Journal of Lesbian and Gay Studies*, 14(2). muse.jhu.edu/article/241316.

Trotz, Alissa. 2007. "Going Global? Transnationality, Women/Gender Studies and Lessons from the Caribbean." *Caribbean Review of Gender Studies*, (1) (April 2007). http://sta.uwi.edu/crgs/abstracts/AlissaTrotz_Going_Global_pm%20_2.pdf [accessed September 1, 2022].

Weheliye, Alexander G. 2014. *Habeas Viscus Racializing Assemblages Biopolitics and Black Feminist Theories of the Human*. Durham; London: Duke University Press. https://www.jstor.org/stable/j.ctv11smq79.

Wallace, Alicia. 2017. "Policymaking in a 'Christian nation': women's and LGBT+ rights in The Bahamas' 2016 referendum." *Gender & Development*, 25(1): 69–83. doi: 10.1080/13552074.2017.1286802

Warne, Peters Rebecca. 2020. *Implementing Inequality: The Invisible Labor of International Development*. New Brunswick: Rutgers University Press. muse.jhu.edu/book/71756.

Wieringa, Saskia. 2002. "Essentialism Versus Constructivism: Time for a Rapprochement?" In *Gendered Realities: Essays in Caribbean Feminist Thought*, ed. Patricia Mohammed, 3–21. Kingston: UWI Press.

Wynter, Sylvia. 2018. "Beyond Liberal and Marxist Leninist Feminisms: Towards an Autonomous Frame of Reference." *The CLR James Journal*, 24(1/2): 31–56. https://www.jstor.org/stable/26752176.

Wynter, Sylvia. 1990. "Afterword: Beyond Miranda's Meanings: Un/Silencing the 'Demonic Ground' of Caliban's 'Woman'." In *Out of the Kumbla: Caribbean Women and Literature*, eds. Carole Boyce Davies and Elaine Savory Fido, 354–72. Trenton: Africa World Press.

2 The Geography of Healing at the End of the World

Black Scholar Practitioners Who Evoke Toni Morrison's The Clearing

Kimberly Nicole Williams

Introduction

Toni Morrison's renowned text *Beloved* accounts a phantasmal nexus of slavery and trauma. Margaret Garner, the enslaved mother who killed her daughter to ensure her child's freedom, is a story that attests to the genealogical wounding that precursors the DNA before Black actualization. Garner is immortalized through *Beloved's* protagonist Sethe, the grieving and recovering survivor. Sethe is not alone in her testimony of confrontation and psychosomatic healing. Baby Suggs, Sethe's mother-in-law, is integral to the development of reimagination that was formerly dangerous for the enslaved community. Suggs creates and leads a spiritual service for an intentional space called The Clearing. This space becomes supernatural with its focus on resurrecting one's body through mindfulness and intimacy.

> When warm weather came, Baby Suggs, holy, followed by every black man, woman and child who could make it through, took her great heart to the Clearing—a wide-open place cut deep in the woods nobody knew for what at the end of a path known only to deer and whoever cleared the land in the first place. In the heat of every Saturday afternoon, she sat in the clearing while the people waited among the trees.
>
> (1987, 88)

Baby Suggs offers her community a new relationship that deems Black skin as natural, wondrous and in need of tenderness. Moreover, Suggs implicates their geographies as sutured to healing, which is important considering the often grisly history of botany and Black death. Suggs widens the capability of the forest—past turpentine labor or reaper of lynching. She deems the landscape Black and comely because she awakens the prayers hiding in trees. This is where Suggs brings her lush of sensory awakening:

> "Here," she said, "in this here place, we flesh; flesh that weeps, laughs; flesh that dances on bare feet in grass. Love it. Love it hard. Yonder they do not love your flesh. They despise it. They don't love your eyes; they'd

DOI: 10.4324/9781003143550-4

> just as soon pick em out. No more do they love the skin on your back. Yonder they flay it. And O my people they do not love your hands. Those they only use, tie, bind, chop off and leave empty. Love your hands! Love them. Raise them up and kiss them. Touch others with them, pat them together, stroke them on your face 'cause they don't love that either. You got to love it, you!"
>
> (1987, 88)

Suggs evokes a particular type of necromancy that awakens their muscular and respiratory system that was previously in inertia from slavery. She offers the community, especially protagonists Sethe and Denver, a becoming into love predicated on spirituality, vulnerability, and redefining familial norms. Accordingly, she summons the community into a circle to evoke their vulnerability through specific tasks like dancing. Moreover, Suggs draws their interiority or their quiet—that plush part of themselves they often had to snuff out in order to survive during slavery (Quashie 2021). Baby Suggs ushers them into redefining love during their new, worldly transition into what Rinaldo Walcott would deem, a long *emancipated time* or "the continuation of the judicial and legislative status of Black nonbeing" (2021, 3).

Walcott further expands his theory of identity and placement with time: "What emancipation does not do is to make a sharp and necessary break with social relations that underpin slavery" (2021, 3). This means that Baby Suggs and the Black Ohioan community are *legally* free but are fixed in the shadow of slavery that articulate new challenges of trauma and fugitivity. This language relates to Frank Wilderson's theory of AfroPessimism:

> Blackness and Slaveness are inextricably bound in such a way that whereas Slaveness can be separated from Blackness, Blackness cannot exist as other than Slaveness. There is no world without Blacks, yet there are no Blacks who are in the world.
>
> (2021, 41)

However, Walcott also articulates the Clearing as the first brink toward freedom in a world machinated on anti-Blackness—that Suggs helps her community to breathe in a new world amidst a grieving one:

> The reclaiming of the flesh as a body, a body loved, is a glimpse of freedom in its kinetic form where freedom meets love, and where love becomes an activating force toward a potential freedom. In this instance, freedom exists beyond the material even though it is also, and importantly, material conditions. Indeed, love, a nonmaterial condition, becomes a major context for moving toward freedom. Black life points us toward what freedom might be, and ultimately is, a project yet to come.
>
> (2021, 5)

The Clearing provides a conduit for a reawakening of the flesh onto the spirit and onto *liveliness* and *living* (Quashie 2021). This "glimpse of freedom" is also a synapse of redefining selfhood and the very definition of love as duplicitous—a space where formerly enslaved people first defined love through their singular being but also with each other. This intentional simultaneity of loving and liveliness through oneness and kinship is pivotal to actualizing Black being. McKittrick notes this specialized kindred:

> Friendship is hard freedom. Maybe friendships effectuate consciousness and liberation and possibility. How has this world already been radically reimagined by and through black thought, and how do we share that knowledge with each other so that we might practice a humanism made to the measure of the world?
>
> (2021, 73–74)

However, this tenderness of collective *and* loneliness precludes the daringness of a love practice in an atmosphere writhing in pain. Quashie proclaims this loneliness as an element of ownership and deliberateness shown through Morrison's *Sula*, the essay, and Black poetics in his most recent manuscript on aliveness: "'I am' constitutes instances of beholding the immanence and transcendence of being… In a black world, one can be of relational oneness, relation as a world of one's becoming that includes being more than one" (2021, 32). This ancestral recall that Suggs curates incorporates a similar restorative bridge between oneness and community. By placing the Clearing in nature, Suggs further parallels this relationship with meditative and paradoxical geographies through its relationship with nature, God, loneliness, and ancestry gatherings.

Black people carry a drop of sun (Fanon 2008) and this means they will continue to provide a lovely warmth for each other and supernatural intelligence to the world, even as their shadows project the Atlantic ship. Because whiteness and its many amalgamations will continue to contort Black muscularity, it is integral to create a space of cultural care. COVID-19 and police killings exhaust the day, disrupting biology, spirit, and time. In fact, a recent study purports that witnessing unarmed police killings results in somatic effects that are similar to a diagnosis of untreated diabetes (Pazzanese 2021). I wanted to study the liberatory potential and theory of the Clearing—more specifically—examining the slippage from page onto earth—especially now, in this crux of witnessing death over and over again. Urgently, I needed a reminder of faith in the exhale that was evoked through interviewing and communing with specific kindred workers who commence The Clearing in real time.

In the tradition of Suggs and Morrison, I sought those who created spaces of love through the interiority and circuitry of nature, expulsion, and movement. In the tradition of Black feminist ritual, I learned about these powerful people through word-of-mouth. My Ph.D. cohort member Dr. Latoya Scott told me about Tallahesse's Rooted Resistance,[1] a liberatory fitness and wellness program for queer, trans, gender non-conforming, non-binary, gender-queer, and historically untapped populations. Roc, the founder and director of the program, purposefully leads this somatic careship outside—an intentional provision related to identifying with the healing of nature as well as being outside, showing queer bodies as organic and here. Courageously, they coordinate this group while also pursuing a doctorate degree. I also interviewed Nia Wilson, the director of the SpiritHouse, after researching its mission from a discussion with healing practitioner Dr. Della Mosely. This house serves as a Carolinan, "multigenerational Black women-led cultural organizing tribe with a rich legacy of using art, culture, and media to support the empowerment and transformation of communities most impacted by racism, poverty, gender inequity, criminalization and incarceration."[2] Lastly, I talked with Dr. Alexis Pauline-Gumbs whose most recent transdisciplinary text intertwines marine biology and Black feminism. I actually met her and have devotedly followed after a creative, feminism workshop she coordinated during my M.F.A. time in 2010. Her previous work includes a wealth of scholarship that intertwines Black feminism, marine mammalogy, creative writing, and more. We talked about her program Sunday Service, which draws on:

> Their [Pauline-Gumbs and Sangodare] traditions and vast range of spiritual and religious traditions that they studied from around the world and weave a coherent fabric that makes a prismatic congregation all feel at home and reflected even as it centers queer identities Black, Indigenous and people of color.[3]

"The act of sharing stories *is* theory and methodology" (McKittrick 2021, 73). These Clearing practitioners actualize Morrison's work into the now. Also, in the Black feminist tradition of oral history and cultural storytelling, I wanted to listen and archive the descendants of Baby Suggs who have dreamed a landscape of impossibility into a deliberate space. I convened with all of them during the height of COVID-19 and police terror when cities were shut down and masks were scarce. I am grateful for their slippage into topics of woe and healing in the conundrum that is Black personhood. *This is the flesh I'm talking about here. Flesh that needs to be loved"* (Morrison 1987, 88).

Spirithouse, Inc.: Interview with Nia Wilson[4]

> *And so in our healing, we're not only healing ourselves in this moment, but we're healing our past; more our future. —*
>
> Nia Wilson

Figure 2.1 The SpiritHouse founders and staff photo. Center: Director Nia Wilson. Photo credit: Jovan Julian.

Kimberly Nicole Williams: **I would love to learn about the genesis of the SpiritHouse. Can you share the backstory?**

Nia Wilson: It was founded by a group of artists and students at Duke University. These included artists from the Mary Lou Williams Center with students and other local artists who later brought poet and artist Amiri Baraka. He actually is the founder of the very first SpiritHouse in Newark, New Jersey. They asked for permission to use the name SpiritHouse when they brought him and then the legacy continued. There's one [SpiritHouse] that was created by activist Ruby Sales too.

The House was always centered on using arts and culture in the Black community and throughout the years, the center became more empowered and rooted in who and what we know, and who we are. We spiritually and intuitively found our voice to continue to heal within and with others from the harms of capitalism and white supremacy.

KNW: **You briefly introduced the meaning, but can you speak to the definition of healing as it relates to your work in the SpiritHouse?**

NW: We carry trauma and it's interesting that science has finally caught up. I mean, science is finally caught up with what we've known our entire lives: traumas that our people have endured for generations are actually carried in our DNA.

And we may not have the words or terms for some of the things that we have been born with but it does exist there—and then we have a layer on top of that. Then, we have to deal with trauma in our own lifetimes on a daily basis when it comes to oppression and you know… racism.

We are being told to fight and we are being encouraged to fight, to push, to force, to make change, but we don't have a lot of conversations about the fact that we are traumatized people who have never had the ability or the space to speak about our daily and ancestral trauma. Even if we win the battle, we have not addressed the trauma that we have endured. We have not addressed what's in our bodies, and so as we continue to build, we build in an unhealed space. We have to remember that most of us who do harm, are those who have been harmed.

That's not natural. That's not our job—to endure. We have to heal and to stop and change our DNA. We have to change what we pass onto our children and our seven generations and beyond. I want our future generations to heal and avoid trauma as a livelihood. We need to say that and speak that incessantly. Healing and therapy was seen as a luxury or seen as a white luxury, but that's changing.

Our ancestors have endured a lot and we should be proud of our people, but it's not natural to continue to take and endure such harm because we can. We have to begin healing and change our DNA.

KNW: **Oh dear, now I'm in a daze as I think about my own healing while being in an academic environment. There's no healing in this space.**

NW: I would suggest you read everything by Dr. Alexis Pauline-Gumbs! Her work is about that and she is part of SpiritHouse too. She was intentional about building a community of care because she understood the trauma Black folks endure and continue to endure in the academy.

KNW: **Yes, I love her work and will interview her as well! I love the connection and gratitude that you extend. I love to study and inquire. I love the verb of being a student but yes, it's something I have to process.**

You've already dipped into these waters with your discussion of healing, but can you also extend your discussion about resurrection too?

NW: Hmm, well we are ancestors. You can look in the mirror and you look like your grandmother; I was joking with my sister today that one of my toes is like my dad's toe. And the scientific fact is that's who we are. So again, as I said, if we are with the physical shape of our ancestors, then we can understand or we need to understand that we are born with their emotional and spiritual consciousness too.

Historically, we weren't able to honor our ancestors, our dead. You know, we don't know so much of our history. We can't count ourselves 10 generations back. We don't know the names, but we know that we physically carry them with us, and so the idea is that as we heal, we're doing it in service of them too—those who chose to survive so that we could be here. None of us got here by ourselves.

Even if you don't have children, you share DNA with other family members. There is someone who is also holding the same DNA that you hold. When I think of resurrection, I think of how cyclical it is—how the past, present, and future are connected at all times. There's no separation.

And we're more than just this moment; we're more than just this time we are here. We carry more than that.

KNW: **Those who chose to survive yes—and those who are surviving. Tell me more about who inspires you and informs your work—from ritual to practice. This also includes songs, works, books, poetry etc.**

NW: Honestly, Alexis [Gumbs] because she created spaces in SpiritHouse that didn't exist before. She challenged me to believe that I was worthy of healing. When SpiritHouse was founded, healing was not part of the lexicon. It was about endurance.

There was a challenge from the younger collective that said, "No, we are not doing it that way. We are more than enduring." This includes Alexis, Adrienne Maree Brown, Ebony Noel Golden, June Jordan, Audre Lorde, and of course more. All of these artists I can read and weep with. All of their work is still relevant in my bones. Black women artists and previous members like Ebony Noel Golden, who has shaped the SpiritHouse and does theater work, have impacted me in this work. We have a former founder, Omisade Burney-Scott, who does research on Black women and menopause. She is also an Oshun priest in the Ifá Yoruba tradition and so we have an indigenous, African spirituality that shapes our work with remembrance and healing.

KNW: **I'm looking at the SpiritHouse website and you all have so many wondrous spaces for repair. I see workshops, interventions for mourning, theater companies—there's a deliberateness in this work. Can you share some of the highlights?**

NW: We do a lot of work because we recognize our lane is our village. One of the most impactful experiences is when we go deep into healing. One summer we went to the river in Durham, and invited folks to come to eat, play games, and we sang as we walked. We got in the river with children and elders and just yelled. We released. It wasn't facilitated. If there was something they needed to release in the water, we were there to support that. If they just wanted to play, we supported that too.

It's a heavy lift. We have a lot of Black people that are beginning to realize they have trauma. It's a heavy lift when 50 people come for guidance and for release, but we also need to release.

We did that three or four weeks in a row and found out that was too much for us. We try to partner in order to think about capacity.

KNW: **Wow, that sounds straight from *Beloved*.**

NW: Yes! When Baby Suggs took them out and held them.

KNW: **And how has this time of social distance and the pandemic changed the SpiritHouse innerworkings and intimacy?**

NW: Again, this goes back to us taking care of ourselves. There was a lot of expectation. There's a lot happening. You know, there's a million webinars. But we're listening and watching and doing what we can to care for one another.

We had local businesses from African and Latinx communities help us with masks. We sought out a printing company that made custom masks in English and Spanish that said: "Because I'm essential." We handed those out because we wanted to take away the capitalist nature of essentialism. We are essential to our families. We are essential to people who love us. And so we wanted to shift that conversation to say that we're all essential and therefore, because we're all essential we all need to take care of ourselves, and we all need to protect ourselves. We all need to be protected.

We did a little bit of a campaign called *Black People Breathe* around everything happening with Black Lives Matter. We deserve our breath. We deserve our life. We deserve our love and we deserve our healing. We deserve to see our value, past what we can produce. This moment means that we have the right to be still and to be healed. We are trying to get Black people to embody that; we have been conditioned to name our value as producers and endurers. That is harmful, so we are still trying to find a new way and shift our values and deconstruct or decolonize our minds and bodies.

I mean now we have people knocking on our doors saying, "we need healing," but the staff and myself need healing too and rest. So, we are doing something new called, t*rusting the ripple*—like that ripple for small businesses that created masks, medicines, and prepared meals. We are counting on the collective ripple for each other. We are counting on community support.

We don't want to exhaust a few people. We have to unlearn that. That's not what our ancestors want and we want to model something different.

KNW: **When I think about healing work, I think about nonbinary, queer, and Black women. How do you think about gender and healing—especially how it relates to Black men?**

NW: Well, we've got a couple of Black male identified healers in our tribe; one of our young men calls himself the East Wizard, and he can name and tell you the properties of any and every crystal that exists [laughs]. Although we understand gender is fluid, we understand that they are multiple identities. We realize traumas and experiences are influenced by gender presentation too. And this is important and pretty relevant because there's not enough Black men in healing work.

When SpiritHouse was created, it was very male heavy, but oftentimes in our community when women rise up, the male leadership goes away. That's been very hard and I have tried over the years to have those hard conversations with men in my life about that.

Our relationships, our communities are not going to heal, if we don't address gender issues. We need more male-identified Black healers and our community to be able to create a space where Black male identified people feel safe to talk about things.

I have a son and you know his father's not around, and when he was younger I used to try to get other Black men to be more present in his life. And you know, I would say to them, you know, "I'm not a man. There's things he's going to experience and I don't have the answers, no matter how amazing you think I am, you know." And I think for a lot of Black men—they didn't know what to do because it hadn't been done for them.

But I see small windows of healing as victories. Yes, I see young Black men loving on their sons in ways that I hadn't seen as I was coming up. And so I definitely see that it's happening with folks who are maybe not even titling themselves as healers or artists.

But then I'm still talking about the binary of male and female. We also have to include the importance of gender fluidity. We have several people in our tribe who identify as queer or [gender] non-conforming. It's important that we can keep our people safe while challenging folks who have been conditioned that their identities aren't true.

KNW: **Yes, conditioning affects that collective love and beliefs.**

NW: There's a young man who was a part of our tribe, who had spent some time in prison. Sadly, he passed away a few years ago, but he said when he came to SpiritHouse, he knew that he had the right to heal. He had done so much harm in the community and being in and out of prison as a younger man. And so, you know, he was conditioned to believe that he didn't deserve healing.

It was important for him to have that conversation around being someone who caused harm, but also someone who endured harm and someone who deserves to unearth that harm. He had the right to heal and he did. And then became a healer himself.

Right, so that the whole thing that I said earlier about *endurance* and you got to think about that too. You know, we are taught to endure and that we're stronger, and that we have to endure. And that belief is embodied. That belief was in us before we were born.

I would suggest reading a book called *The Healers*; it's one of my favorite books. It emphasizes how we form. As I said earlier, healing comes with cyclical cycles of the past, present, and future. We're not going to shift the DNA of that in a year, but we will see small windows of healing as the victories.

And then it doesn't have to happen with someone who is identified as a healer because they can be artists; they can be you.

Rooted Resistance: Interview with Roc[5]

I didn't create this because of my wound. I created this because I felt encouraged to have this.

—Roc

Figure 2.2 Rooted Resistance camp in action. Roc is pictured in the center with the right leg lifted. Credit: Ronan Wilson

Kimberly Nicole Williams: **Talk to me about Rooted Resistance. Tell more about the genesis of this program.**

Roc: Throughout my body are histories and legacies of slavery and colonialism. Throughout the land are histories of violence and erasure and dispossession. This includes indigenous people. I mean, at its core, dispossession and power is how sports came to be. They are tethered to sports and basketball arenas. In fact, all of this land is attached to erasure of indigenous people and formerly enslaved, African people and so moving our trans bodies, our queer bodies, our Black bodies … all gets placed in a political space.

Rooted Resistance started as a "let's work out with the homies" outside of the institution where you visually never see yourself or have to be questioned or interrogated. I mean, we all have stories and we all heard the same questions or at least felt it: "Is that a man or a woman?" We were merely trying to release and cultivate our health and bodies, but we were also placed in terrorizing spaces.

We're entering a fitness center to expand our health and exercise our heart, to exercise and flex muscle and our endurance and our stress and some body weight. So, why would we subject ourselves to the social institutions that demonize us? I guess that's a part of the foregrounding of it. We workout outside to honor the shared, sacredness of the land but to also represent that visibility.

KNW: **That's incredible, and I think there're so many key points that conjure so many thoughts—can you speak a little bit more about the intentionality about land and working out outside?**

R: Yes of course, the intentionality is connected to all my own questions, my own curiosity, my own undoing, and learning around space and place. Being *outside* constitutes a reciprocal relationship—we don't own the space. What we're doing is expanding them and expanding what we think they can be and completely refusing the social relationships, the way that we operate in a condition of cooperation. I read something before where it said the earth is an ancestor.

The park where we work out is not something we own. We are expanding that social relationship of what place can be and we want to rethink our relationship to each other and ourselves—that's behind being outdoors.

The other side is around access and financial access to a gym membership. A membership is money and many communities are dealing with limited funding and scarce resources. There's transportation and things that are associated with poverty, especially for trans people and queer people. When you think about housing, finances, and unemployment, there are barriers to our existence in this racist, capitalist society. The LGBTQ+ community, I mean we have barriers to those access. We want the ability for people to show up and avoid financial transactions. I think maybe people would come more and feel less shame about deficits and also wellness, if money is not a contingent factor.

KNW: **Okay, so this makes me think about this undoing with shame, body, place, and access. Tell me more. What does healing mean to you with your work?**

R: I can answer with just words and draw it, but healing feels like belonging but belonging to myself first.

That's what I feel and that I want to be really rooted with healing in as much as an individual process as collective. In order for me to be part of a collective, I have to invest in myself with intentionality and curiosity.

You know, we are the ones that are creating and we are the ones that are moving. This is an evolution and revolution. Someone told me that Rooted Resistance is an alternative healing story, but I wouldn't say that because alternative means that I am comparing this [Rooted Resistance] to the majority institution. Instead, I would say this is necessary and imperative, healing work.

More specifically, I didn't enter Rooted Resistance thinking about trauma because it's also about joy. It's about how I felt moving throughout my body through sports or playing outside with my parents at our basketball hoop—what freedom I felt soaring in the air with basketball or rollerblading. In fact, I remember how my parents would have to yank me away from outside at night, but they always encouraged me to play.

I didn't create this because of my wound. I created this because I felt encouraged to have this [Rooted Resistance]. My mother wanted us to be dancers. My sister can draw her butt off and we even participated in theater. We were afforded opportunities to appreciate the arts and the body. But I pretty much can only draw stick figures and I didn't enter the dance world. I just can't do a split [laughs]. But dancing is so important and sharing songs too! When I think about Rooted Resistance, we are flailing our arms and so many feelings in our body. I am actually learning more. This is about the embodiment.

I want to do *trauma-informed* work. You know, back in the day when I got my personal trainer certification, the person who was leading it was demonstrating the exercise by touch. I'm like, "you don't need to touch anyone" without permission. They were telling me this demonstration was trauma-informed, but we have to talk about consent and the history of the body. With the fitness industry, it's commodified and highly commercialized, so all these buzzwords, just like social justice, world trauma, trauma-informed, this is deeply serious work. It's not a static practice.

Transmasculine people and other trans folks, we have very similar stories, traumatic experiences, interpersonal relationships, romantic relationships in this world. This is just by way of the structural hierarchies and binary foundations of colonialism; we are impacted by that on different levels. When I was 15, I didn't know that language. You know, when I was a teenager, I mean, I didn't use words like colonialism, racist, or sexist, but I felt racialized being Black in the U.S. That was something that was very clear to me about the neighborhood and society. But I didn't understand the concepts like colonialism and terrorism and how deeply ingrained all of these things are.

So that's where I'm coming from you know—digging into what it is for me, the geographical land, and the ears of the communities.

KNW: **How does the concept of faith and resurrection align with your work?**

R: I don't use those words in my work but they have meaning. Those words are defined through memory. When I think about faith, I think about the complex relationship between religion and spirituality with my community. But if I separate and examine faith, I think about my mother and grandmother and it's something stronger but that's not how I speak.

When I hear resurrect, I get goosebumps, I think of unsetting and unearthing. We are going to dig and triumph. I think about faith and resurrection as a pendulum of resistance and liberation. Resurrection to me? Well, I don't think you can rise up without destruction. Something has to be destroyed.

People have come to our camp and have said, "Roc, you're taking us to church." Like, I know what that means with choir, rhythm, and the space of family. But going to church is not liberating with everyone. I grew up Catholic and stopped going at 15. I wasn't a faithful person going to church; I felt that I was repressing myself, but I certainly think of blackness when I hear those words [resurrection and healing].

KNW: **I like how you mentioned play and conjuring. What are some other places, texts, people, do you draw your well from?**

R: Caroyln Finney in Black geography with her text *Black Faces, White Spaces*. I was like is she in my brain? There's so much power in how she bridges critical race theory and Black studies in how dominant narratives influence belonging in the environment. There is an intentional history of who belongs as a commodity and leisure. It's incredible. She also evokes collective memory and that is something I want to dive into more. I would certainly say bell hooks too. Actually, I have to read you this quote too; it's from the *Sisters of Yam*,

> In modern society, there is also a tendency to see no correlation between the struggle for collective black self-recovery and ecological movements that seek to restore balance to the planet by changing our relationship to nature and to natural resources. Unmindful of our history of living harmoniously on the land, many contemporary black folks see no value in supporting ecological movements, or see ecology and the struggle to end racism as competing concerns. Recalling the legacy of our ancestors who knew that the way we regard land and nature will determine the level of our

> self-regard, black people must reclaim a spiritual legacy where we connect our well-being to the well-being of the earth. This is a necessary dimension of healing.

I mean that quote is so moving because she is locating the space between us, land, and healing. Paulo Fierrie, *Pedagogy of the Oppressed* is also a big one. I always go back to that book. There's always something new that I didn't see. There's a handful of geographers like Ruth Wilson Gilmore and Katherine McKittrick. They so pointedly name prison as a racial geography. I would say all of their work informs my research as well as critical sports theorists. They help me make connections and bridge community organizing with theory.

I also think community organizing bridges that gap between the academy and the community. Black Marxism has been richly important for me and studying race before feudalism. But I can't end without saying Staurt Hall too. His purview on cultural studies and West Indies scholarship—I mean his activism has really helped me pull in this conjunction where time, linearity, and refusal meet. All of his work cycles through Rooted Resistance.

And of course my family—my parents and grandparents and their experiences with labor, the garden, and land in Louisiana played a considerable influence on geographical and bodily care.

KNW: **What are you seeing next up for Rooted Resistance? What does futurism look like for Rooted Resistance?**

R: I'm not really keen on having a storefront and that's not part of my vision, but I want to have people connect with it everywhere. The virtual world has offered that space. I don't know where Rooted Resistance will be but I want it to have quarterly events like rock climbing, somatic retreat, and a yearly larger event. Also, for my own personal growth, I want to create somatic workshops and be able to expand. I hosted a rock climbing event where we learned together.

KNW: **Can you describe a memory that was impactful during your time with Rooted Resistance.**

R: There have been plenty of times, especially when I first started because I was the only one who showed up for a Rooted Resistance event. I understand that because I was a new member of the community and I was still marketing and meeting with people, but I remember this one time, I was really angry with the political social climate and recent policies with the military, so, I was like, I'm going to have a Root Camp! There was this wonderful counseling office that even marketed the event. These two participants came really late but I told them, next time. That meant a lot to me, being so fresh and new that people came to build together, however late, that the love was there.

Sunday Service: Interview with Alexis Pauline-Gumbs[6]

I mean this is something my dad would say, "A caterpillar running faster and faster and faster is not going further and further and further—it's disintegrating and becoming something else; this butterfly is like existing and flying as far as it can fly, but only for a short time."

—Dr. Alexis Pauline-Gumbs

Figure 2.3 Sangodare and Alexis Pauline-Gumbs in a virtual Sunday Service. Credit: Courtney Reid-Eaton

Kimberly Nicole Williams: **Can you tell me about the genesis of Sunday Service?**

Dr. Alexis Pauline-Gumbs: Yes, it was really created by my partner Sangodare who is so amazing (but I guess I'm biased and yet I'm pretty sure about that). Sangodare went to divinity school and has a background in film and art. But it was actually here in Durham when we participated in a sermon slam in honor of Pauli Murray that Sangodare decided to preach a queer, Black feminist sermon and it was like ahhhhh! I mean, it was everything! Myself and others were memorized and because of that, we were invited by Northstar Church of the Arts. Now, we originally had it planned for every Sunday, but then scaled back and really thought it was important to continue because of the many tragedies in 2020.

We had this grounding space every Sunday to draw on the divine wisdom of Black feminism and share it with our Community during such a turbulent time.

KNW: **How has Sunday Service provided a space of reflection and growth?**

APG: It's an energy shift in process. Sangodare, in real time, is actually going through an energy shift while the sermon happens. This year, it's been just me in the room. People make fun of me because it sounds like I'm the choir, the amen, and the deacon board. This year has been so intimate and I have grown from the sermons.

I literally get to feel and actualize Black Feminism energies shift through Sangodare's sermon. I've grown from experiencing those sermons, but I've also grown from sharing how these sacred texts—from Lucille Clifton's works to Audre Lorde poetry—how are they asking me to grow? I've shared my own poetry, unpublished works, and real-time experiences. I've cried on camera. It's been an opportunity to grow in my vulnerability.

KNW: **Can you talk about the influence of the pandemic and the election[7] on Sunday Service? How has this space been a balm?**

APG: We are really committed to being responsive in our communities. We are also committed to providing an alternative to anxiety-producing feedback from social media, especially with the constrictions that people have. The pandemic has a huge impact especially on Black lives. The ongoing police murders on Black people and accountability. It's like a toxic shock. The actual physical, emotional, and spiritual impact of having a white supremacist head of state is too much.

We are not trying to say the smartest thing; we are trying to show that in the face of everything, we have the divine in us and to return to love and channel love.

There is something cleansing about loving ourselves intentionally. It's always necessary but especially so when there's so much disrupting that. It's definitely been a balm and we have definitely been in Gilead. We realize there's a need for our service, but we feel so supported by this whole congregation of people that are channeling love and remembering that it exists.

That itself is faith-building. There's critically connected, critically masses of people who in that moment, channel love. It has allowed me to center love as the basis of my decision making.

KNW: **I think about 2020—the inescapability of trauma, lineage, family—it was a difficult year but also a space to be intentional. I could not escape trauma and my selfhood. It was difficult but maybe a conduit for disruption. Your ability to create multiple Morrison-mused Clearings with the Black Feminist Relationship workshop and then the Black Feminism Filmmaking Collective and your Meditative Oracle Workshops is astounding. And then you still have Sunday Service too! The dexterity**

and vulnerability—the supernatural quality of your creation is so impactful. How do you refuel? How do you fill your cup?

APG: There's a show called *Woman*, like this feminist public access show. Marge Piercy and Audre Lorde are on it. During one of the episodes, the hosts ask about synergy, audience, and that relationship to readings. Lorde talks about how poetry touches that emotional core. In live readings, it's really about touching that emotional strength of an audience and that interchange of energy is how Lorde got her fuel to write more poetry. I was thinking about how Audre Lorde would answer that, so that came up.

For me, I do feel nourished and fed when people respond. I get energized by all the love sent via zoom room messages or insta messages. It's very energizing but we still have to eat, watch a movie, and clear space.

We understand it's important to really rest. We honor the fact that we are holding space for all of this energy coming through. It's not a sacrifice and rest remains important. Daily facetimes with my nieces are so nourishing. I think intentional rest and also just love and play between Sangodare and I and our family, is really what helps us. But the ongoing practice is pivotal too. We know our spiritual connectedness is not disconnected from the collective. And we have to continually put into our process if we continually share. We wake up very early to have deliberate, ritual time. No one wants to have meetings at that time. That's very, very important. We balance and process honor and being honest with ourselves.

But we realize this work is soul work, and it's changing us and so we need a cycle like a butterfly. We are in the process of creating and creating accountability. There has to be a butterfly rotation. There has to be chrysalis and being okay with being a caterpillar. Like we understand, we are transformed by our community and created with accountability. I mean this is something my dad would say, "A caterpillar running faster and faster and faster is not further and further and further—it's disintegrating and becoming something else; this butterfly is like existing and flying as far as it can fly, but also for a short time."

KNW: **Can you speak more about the relationship between resurrection and rebirth with your work, particularly as you discuss that butterfly metaphor?**

APG: This is the thing. I'm writing about Audre Lorde right now and she thinks about death and resurrection a lot so this is all up in my head. I'm in the process of memorizing one of her favorite poems, *Renascence* by Edna St. Vincent Millay. So, all this is like ding ding ding! I think my own discussion with grief and death relates to my kinship with my father. There's something that is very core to some work with him; there's listening, poetic processing, images, and a healing movement. In my own ancestral practice, I'm in ceremony with my ancestors. I don't know if I would like to call this astral travel or visualization, but I'm in an intentional space with them. That's not me resurrecting them, but a relationship to life and energy that is something that can always have access and continue to exist.

When I think about what it means to have a kinship with people in the present who have died, that is Black resurrecting technology. It's not something that happens once. It's not linear. It's a spiraling of presence. This is how my father thought about healing. You heal it for you; you heal it for all of us. It's forward and backwards. My life as a resurrection is clear to me and is not linear. My presence is the living of many who have died and our energy lives on beyond our lifetimes.

I think another spiritual question to ask is how to be accountable to that and how to remember that—especially in a capitalist-materialist structure, where I'm not taught to make decisions from that ancestral knowledge and intuition. But I know that and I feel that.

I think about these Black feminist texts as oracle and ancestral texts. It's a participation in eternal life. These people are still impacting us. They are part of our 'today' and those possibilities. I mean who they are, shifts with us too! We continue to shift and so their lives aren't over. I mean rebirthing resonates but what really resonates with me is this process of *eternity*.

I wouldn't say we are bringing back Black feminist theorists from the dead, but rather it's eternal and they are still living. As the Combahee River Collective said on a pin: *Black Feminism Lives*. That is the Easter message. Black Feminism Lives.

KNW: **Okay so speaking of eternity and muses, and because you mention the Combahee River Collective, I really want to learn more about your relationship and learning from water. Your most recent text[8] explores so much with yourself, tenderness, sound, and breath. Oh and I love your sound piece with Sangodare on your site that accompanies this text! I would love to hear more about these relations of rhythm, water, and just singing.**

APG: I am so astounded at Sangodore's work as a composer and as a sound healer in training. I'm grateful for how much I've learned from Sangodore. It's essential that what we do is vibrational. We are vibrating constellations of energy. With Sunday Service, we are aware that there is something that changes us, sonically, with breathing and chanting. I mean, it's also changing us in real time. It's transformative—whether reciting Lucille Clifton's poem or chanting a Black feminist breathing mantra—we understand that sound changes us. This is energy work and sound shapes energy.

But Sangodore often composes something from the oracle and also something from Sangodore's childhood to where there is moaning and energies, very much from the Carolinian, Black Baptist tradition and then I'm transformed! Then I get up and do the oracle and then I'm transformed, and then Sangodore is like let's do some singing and I'm the choir! And then we are also laughing too! That's the key thing: we are sound. Sound is so powerful. We allow ourselves to be transformed through sound. I think sound offers an audible surrender.

I guess I would point to our Father's Day service as an example too. My father is an ancestor and I knew I wanted to use his book as a sacred text. I didn't even give myself space to think about how emotional that would be. It was certainly the case that the insights and things I've said during that sermon, I wouldn't have said those things out loud about Father's Day from a Black feminist perspective. I think it's so important to hold sacred space for my community, and it tricks me into holding space for myself that I also certainly need and deserve. I don't understand how much I need it before I'm actually experiencing the sermon.

KNW: **I'm nearby so many tragedies like an hour or two away from the Rosewood site, Newberry Six lynching, Ocoee Tragedy, and St. Augustine that shelters Fort Mose—the first free enslaved settlement. It's amazing because I find myself singing more. And it's related to this quotation I found by St. Augustine, "When you sing, you pray twice." I feel like I have to sing more as a protective measure because there's so much unrest here and sorrow that's intentionally covered up by our Governor and Floridian history. I like to sit and listen to St. Augustine tides and flows.**

APG: I have a cousin who recently Facebooked messaged and inquired, "Lexi, do you sing?" And he has a sister who really *sangs* but I replied, "yeah, I definitely do." He talked about how much he hums when he tunes his cello. But it was really helpful that he asked me that question today. I feel like we need more singing. Singing is remembrance.

We're Jamaican and my grandmother actually called us back to that place [St. Augustine] to meet with family; it was actually 3 generations of family. It was so memorable and extraordinary, and you're right, St. Augustine is so deeply, intentionally shrouded. St. Augustine, like all waters, has so many energies—I mean the shells on the walls and the indigenous linkage—it's so spiritual.

KNW: **I want to end on a memorable experience you had with Sunday Service or one of your workshops. Can you share an anecdote?**

APG: Oh yes—I had a Black Feminist Breathing Course. This was 21 days of a different ancestor and breathing mantra and then an additional 7 days of developing a personal relationship with an ancestor. Now at the follow-up workshop, we were discussing how to listen to our ancestors and seek or look for messages and then suddenly we saw this huge rainbow all over Durham. It was this whole rainbow over the city. It was like "what do the ancestors want to say to us" and then boom! It was so shocking! We made a video of it too. You saw people coming outside to take pictures too.

Oh, and another rainbow happened before when we launched the Combahee River virtual anniversary pilgrimage; we saw it again. We saw this huge rainbow that extended into the steeple of Hayti Heritage Center which extended from the building to the first Black church here in Durham. Sometimes our ancestors like to show out!

It's important to allow yourself to witness an ongoing miracle, whether it's the rainbow or the literal rainbow or whether it's just like the love of all the people who are participating in Sunday Service—it's important to believe in a miracle.

Conclusion

The pandemic pinned us down in omnipresent grief and solitude that altered time: the hours became years and months became days. Moreover, the police and its prison industrial system unearthed Black people into a paradoxical and rigid plantation life. The geographies of anti-Blackness are lively and breathe into every measurement of time and this living embodiment of horror ages Black people faster to the grave. But there's a rhythm on healing that the above interviewees found and this rhythm situates into an altar ritual.

Nia Wilson mentioned how trauma is passed down from generation to generation, but that healing is a tradition too and its cyclical manifesting can create an avian thermal where Black people can trust the air again.

I found several links throughout Roc's, Wilson's, and Gumbs' testimonies:

1 First, they all shared and were adamant that healing was a cycle, and the poetry of this continuous flow, this inertia was startling because that is also the case of plantation geographies. This parallel relationship then gave me a sense of faith because this intersection of time looping felt formidable to counter the vertigo of time travel on the Atlantic. This relationship of centering stillness and trusting the healing movement was also discussed in *Undrowned* (2020)—Dr. Alexis Pauline-Gumbs' most recent text.
2 The discussion of solitude and self-love while also loving the collective was a bridge. Furthermore, this was important as Nia Wilson identified her disruption of "endurance" as key to her evolution of love and healing.
3 All of the healers explored and deemed nature and landscape as kin or respite in their belonging. This is an important relationship to evolve their sense of time and place with nature beyond witness but as an extension of their body and a leap into the unknown.

Finally, their language of surrender in multiple iterations defied the living atmosphere of anti-Blackness, at least for an exhale, and established that in this obstinate holding of violence, that Black people can carve out new worlds in the midst of dying ones.

---For Kiru----

Notes

1 https://www.rootedresistance.com/about
2 https://www.spirithouse-nc.org/#intro
3 https://www.northstardurham.com/eventscal/2020/3/15/sunday-service-w-alexis-pauline-gumbs-amp-sangodare
4 Nia Wilson (of Spirit House) Zoom interview with the author, January 2021.
5 Roc (founder of Root Resistance) telephone interview with the author, January 2021.
6 Alexis Pauline-Gumbs (of Sunday Service) Zoom interview with the author, January 2021.
7 2020 U.S. Presidential elections.
8 Gumbs, Alexis Pauline. 2020. *Undrowned: Black Feminist Lessons from Marine Mammals*. Oakland Ca. AK Press.

References

Fanon, Frantz. 2008. *Black Skin, White Masks*. 7th ed. London: Pluto.

Gumbs, Alexis Pauline. 2020. *Undrowned: Black Feminist Lessons from Marine Mammals*. Oakland, CA: AK Press.

McKittrick, Katherine. 2021. *Dear science and other stories*. Durham: Duke University Press.

Morrison, Toni. 1987. *Beloved*. New York: Randomhouse.

Pazzanese, Christina. 2021. "How unjust police killings damage the mental health of Black Americans." The Harvard Gazette. Harvard University Accessed 17 January 2019. https://news.harvard.edu/gazette/story/2021/05/how-unjust-police-killings-damage-the-mental-health-of-black-americans/.

Quashie, Kevin. 2012. *The Sovereignty of Quiet: Beyond Resistance in Black Culture*. New Jersey: Rutgers University Press.

———. 2021. *Black Aliveness, or A Poetics of Being*. Durham; London: Duke University Press.

Walcott, Rinaldo. 2021. *The Long Emancipation: Moving toward Black Freedom*. Durham; London: Duke University Press.

3 Public Scholarship as B(l)ack Talk

African Feminist Collaborations in the Academy and Online

Rachel Afi Quinn and Maurine Ogonnaya Ogbaa

Prologue

At a large research institution in the South, a second-generation Nigerian American graduate student, Maurine Ogbaa, is writing her dissertation about African feminist theories and African coming-of-age novels and must look beyond the English department for a black female scholar to serve on her committee. With roughly 345 permanent faculty members in the College of Liberal Arts and Social Sciences at the time, only a small fraction are black faculty, and of those, just a few are Africans. Yet Maurine knows of one gender studies professor, Rachel Afi Quinn, who had eagerly attended "(Re)membering Africa: Women's Narratives on the Continent and Beyond," the symposium that Maurine had organized in Spring 2019. The gathering provided scholars with intellectual engagement that they could not find on their campuses. It also brought Zimbabwean author and filmmaker Tsitsi Dangarembga to campus. Over the years, Dangarembga's *Nervous Conditions* (1988) has remained the only African novel Dr. Quinn teaches in her class on black girlhood. Dangarembga's presence brings Dr. Quinn to the symposium in the kind of intimate and dynamic space that Dr. Quinn could not have encountered elsewhere on campus. Later, Maurine approaches the junior faculty member about serving on her dissertation committee, and Dr. Quinn, open to the opportunity, responds by asking Maurine how she thinks about feminist academic work and feminist methodology; this leads to a larger discussion about books, organizations, conferences, ideas, and the difficulty of finding black mentors on campus.

Later, when Dr. Quinn seeks to develop a digital humanities project around the experiences of Black LGBTQ+ migrants in the South, she thinks of Maurine as a graduate student who might be interested in being part of the development team. When an opportunity for internal funding arises, Dr. Quinn hires Maurine to develop the project website and works with her to coordinate the broader digital humanities (DH) working group. Maurine does not have much familiarity with DH or public scholarship, but she has recently learned how to create a website. She comes to the project interested in the topic and is ready to develop new skills. Dr. Quinn imagines these new skills could help this second-generation African student by expanding her training and perspective

DOI: 10.4324/9781003143550-5

around how she might use her Ph.D. But unbeknownst to Dr. Quinn, Maurine has also just discovered that she would be without funding for her final year of dissertation writing, placing her in the precarious position of figuring out how to provide for herself while also finishing her doctoral program. Dr. Quinn had had a similar experience at the end of her doctoral program. Her intervention is an essential support for Maurine at the moment in which black graduate students are often left to sink-or-swim before completing their degrees.

Introduction

This chapter is inspired by Maria C. Lugones and Elizabeth V. Spelman's influential 1983 essay "Have We Got a Theory for You! Feminist Theory, Cultural Imperialism and the Demand for the Woman's Voice," in which they model the power of collaboration across difference, stating:

> We are both the authors of this paper and not just sections of it but we write together without presupposing unity of expression or of experience. So when we speak in unison it means just that-there are two voices and not just one.
>
> (Lugones and Spelman 1983)

This chapter is a dialogue between a graduate student and an untenured professor that interrogates the significance of our presence as African women academics in the US South. Lugones and Spelman take up the form of an ongoing dialogue to explore and define the differences between them as women and feminists—one white/Anglo, one Hispana, one from the US and one from Argentina. In our case, Dr. Quinn is Ghanaian American and Maurine is Nigerian American; as a mentor and mentee, we have gotten to know one another through intellectual collaboration around migration and anti-blackness and our connections to African feminism. Initially, our relationship formed out of necessity and discovery. We found one another as African feminist scholars—minorities within a minority at our institution—wanting to collaborate and learn from one another. We had identified intersecting interests in African literature, coming-of-age texts, and our desire to work on blackness and gender in the academy, but we also recognized our comparable and overlapping racialized experiences within the academy. Thus, we set up and invested in a formalized mentor-mentee relationship as a framework offered by the academy for how we might relate to one another. Throughout this essay, we also interrogate the hierarchy under which we began our relationship in the academy, naming our differences and similarities as African feminist scholars while in dialogue and collaboration.

Like the many black feminist scholars who have cleared some of the path for us in academia, "we affirm and celebrate the idea that black feminist scholars and those who seek to understand the lives of women in Africa and the diaspora have multiple identities and choose to speak in many voices" (Rodriguez,

Tsikata, and Ampofo 2016, ix). This chapter provides us with an opportunity to think together as African feminists in the diaspora, name insights provided by our own positionalities, and specify our relationship to African feminist thought. Woven together are our shifting perspectives. Mentor seeks to educate mentee about what she has learned about structural racism and sexism in academia and the value and effort of public-facing scholarship as black feminist methodology; and mentee, in turn, shares reflections on lessons learned. Such a space of reflection first and foremost allows us to articulate how African feminist thought has been critical to how we navigate the North American academy. Our investments are in the study of experiences of blackness and Africanness, and our intellectual commitments to African feminism and African literature. We also identify some of the challenges we face in attempting to center African feminist thought in our work in gender studies, cultural studies, and Anglophone literary studies. We both claim a black feminist pedagogical praxis, an approach that is itself necessarily in constant dialogue with feminisms shaped by various national, racial, and economic contexts.

As African feminist scholars of the diaspora, we contend with a unique form of institutionalized misogynoir as we carve out spaces for our scholarship in and beyond the academy. Underlying our experiences and the perspectives we offer is the model minority myth applied to us as African women scholars in the academy in North America and the positions of privilege that we hold relative to African feminists on the continent. With this chapter, we explicate our investments in West African feminism as a praxis which shapes our approach to collaborative work. Further, we identify how African feminist knowledge, online/digital engagement, and circulation of theory inspire our collaborative work in concert with other scholars and activists. African feminist genealogies name our fellow troublemakers and trouble us to think about the contemporary issues we encounter. We consider how our collaborative work across disciplines and our public scholarship make it possible for us to imagine an alternative future for black women as intellectual laborers in fields that are steeped in Anglo-American authority. Ultimately, we name the ways that African feminisms are essential to the theoretical frameworks we employ as scholars in the US and are also tools for our survival in the academy.

In Which We Ask Ourselves, "Who Are We to Do This Work?"

> *In the voice of a black, female PhD candidate who entered a program after teaching composition for several years and writing fiction. Also, in the voice of a graduate student who is African sometimes and black other times, depending on who else is in the room.*

I understand the question—Who are *you* to do this work?—to be about my self-identification as a Nigerian, African, and transnational black person. People who encounter the embodied me might wonder: Who is this woman born in Austin, Texas, raised in Houston, and who talks like "that"? What is her investment in African feminism? If the text that is my body—from the brown hues of my skin, to my full lips and nose, to my hair texture, to the difficult-to-describe ways that I carry my body—bespeaks an American blackness, then who am I to espouse an "African" feminism? The answer is that I come to the work of African feminist theorizing and praxis through my own curiosity about radical and transformative thought within Nigeria and throughout West Africa.

The path that led me to African feminisms is littered with stories of irreverent women and smart women, first wives and third wives, and also Catholics and Pentecostals who don't believe a woman should *misbehave*. These are family stories that resist the naming of "feminism," finding instead explanations in "tradition" or allusions to an exceptional or eccentric personality. Moreover, my positionality is informed by my experience as a child of immigrants, a condition which led me to seek out a vernacular with which to understand my own ideas and communicate them to others. I seek, in part, because I was born in 1980s Texas, where the language of the Other was lesser, a liability, an obstacle to "achievement," or so the teachers told my parents. In the State of Texas, the state of blackness, and the state of insecure citizenship—as my household experienced it—English was the language to be learned and spoken. Thus, as I speak English today, I call past into present and contend with the reality that I am not a native speaker of Igbo, though I am a child of people who feel most secure in *their own language*.

In search of a vernacular that I could access and build with others in, I found the work of African female authors, like Buchi Emecheta and Ama Ata Aidoo. I embraced the characters and later the authors as fellow troublemakers. I then began to write creatively, as they did. Finally, I returned to the university to pursue a PhD. My dissertation draws on a broad platform of West African feminism to analyze the various ways that female-authored African novels present the development of adult identity and subjectivity in relation to gender constructs and societal developmental structures (such as the family and non-governmental organizations). My study engages with the work of well-known authors, such as Chimamanda Ngozi Adichie and Aminatta Forna, alongside that of lesser-known novelists, like Amma Darko and Adaobi Tricia Nwaubani. I incorporate African feminism's critique of bourgeois ideals, such as those found in Adichie's novels, as well as its attention to economic and social structures. As I do this, I remain aware of the question: Who am *I* to do the work of African feminism? This question reminds me that I cannot speak *for* women or people more broadly, whether they are on the continent or members of a particular class. Rather, I must draw on my own embodied and material experience, I must avoid universals, and finally, I must engage with feminist theories, which emerge from the continent with the sensitivity and care that black and transnational feminist scholarship demands (Nnaemeka 2003; Steady 2007). With this orientation, I am *just* the right person to do the work of African feminist theorizing, and I join a large group of thinkers. I look to the left; I look to the right—we are many.

In the voice of a black junior faculty member who has had to remind students to refer to her as "Dr. Quinn" because of how quickly they become informal with her. Also, in the voice of a mixed-race African American scholar deliberate about building black community.

As the daughter of a Jewish mother and a Ghanaian father, I have often considered the ways I experience the world differently from my many half-siblings who grew up in West Africa whose framework of race determines me to be white and how differently I experience the world from my Jewish American relatives, who know me to be Black.[1] For both sides of my family, education is a critical marker of achievement, though neither side has a clear understanding of how I approach my role as a black university professor.[2] As a scholar, I have examined the power and usefulness of black feminist theory in teaching critical race theory and looking at transnational iterations of blackness through visual culture and ethnography. I have applied Black feminism as a tool for community-building in public scholarship in the humanities, first in my ethnographic research and writing about Dominican feminists, later as a co-founder of South Asian Youth in Houston Unite (SAYHU), and now with students, faculty, and community members examining the needs and concerns of Black LGBTQ+ migrants within the region. All the while, I have been working to make a place for myself in the academy in which my interdisciplinary scholarship on blackness is legible. Where *I* am legible. I do so among structures that were built to resist inclusivity, and amid hierarchies that have traditionally looked down on "scholar-activism."[3] I work from within the academy to push the door wide-open so that others who value and produce feminist thought might join me. *This is part of my labor.*

I have been trained in transnational feminist theory as a framework for understanding the lives of women across contexts. While I have focused on the experiences of women of the African diaspora, my mentors have often been non-black women: South Asian women, Latinx women, and Arab American women.[4] I have deliberately sought out black women scholars as mentors and co-conspirators, understanding their insights in the academy to be uniquely invaluable to me. I also have found these women to be overextended and in high demand, making our time for meaningful connection fleeting though cherished.

I imagine that my own diasporic identity, queerness, and mixedness make it difficult for Black students at my institution in the South to relate to my experience or draw on the racialized knowledge I offer that differs from how Black identity has been taught in our university's Afrocentric Black studies program. I have also found that a survivalist imperative in the academy among first-generation college students and a sense that "self-reliance as a necessary ideology" (Steady 1981) in the context of anti-blackness makes it challenging for any of us to do collaborative work. These habits of behavior in hostile terrain are reinforced when students and scholars access the academy having built a protective shield of the highest of standards for ourselves, and those standards are often unforgiving. We internalize the idea that it is possible (through perfection or drive) to successfully navigate an anti-black system. But we come to realize that there is no way to do so unscathed. Moreover, behaviors that were once an asset become a liability. Similar to what Maurine has shown me, Nigerian feminist philosopher Nkiru Nzegwu theorized these institutional violences, decades ago. As Nzegwu's example demonstrates, the epistemic hierarchy interpolates white women, in particular, as superior knowledge producers and black women as inferior theorists and scholars. This not only reproduces the racialized and imperial context, it short circuits feminist coalition building, collaboration, and knowledge production (Busia 2003; Oyewùmí 2003). Today, African American historian Koritha Mitchell highlights a long "tradition" of *know-your-place aggression* directed at black people and evident in the academy, through which we learn that we cannot be *too* knowledgeable or *too* vocal if we seek access to spaces not made for us (2018).

Being Model Minority

In the voice of an all-but-dissertation PhD candidate who grew up working class, attended public school, and was teased for being "African." And in the voice of an untenured Ghanaian American scholar who has benefitted from a private school education and proximity to whiteness at a cost. In the overlapping voices of two black scholars who navigate institutions full of people who want us to know that they aren't racist.

Together, we consider: "Who is black?" Our experiences as black scholars in the academy during the first decades of the twenty-first century have been marked by the murders of Sandra Bland, Michael Brown, Breonna Taylor, Atatiana Jefferson, Trayvon Martin, Tamir Rice, George Floyd, and scores of other black people by police. Our experiences are likewise informed by the ongoing assault of black trans women, and the social and popular media discourse tracking this violence with #SayHerName and various dedicated hashtags that have gone global. These daily realities of anti-black violence and police brutality invariably haunt our experiences within the ivory tower; these violences echo in the classroom as we teach a diversity of Texas-educated students who are immersed in a culture of anti-blackness but have no language to talk about it. Many Texas students, African American students included, toe the line by embracing a culture that erases the history of anti-black violence. And black cultural critics like ourselves are forced to adopt the dominant language and make use of it for critique, yet we can miss the mark or be "co-opted" within these systems (Christian 1988, 61). We are endlessly called on to respond to these realities and help non-black students come to an understanding of them, or alternately, sit without remark, withholding any "b(l)ack talk," as Maurine describes it.

I (Maurine) identify as a black, diasporic African woman. I recognize that the opportunity to claim association and belonging in more than one black category is a privilege. The representation of African immigrants and their children as college-educated professionals in science, technology, engineering, and medical (STEM) fields has led to their positioning as a "model minority" (Ukpokodu 2018, 69–96; Anderson 2015). Yet, the model minority trope relies on an uncritical understanding of the factors that create differences in the rates of college enrollment and graduation between African immigrants and their children and non-immigrant US Black people (Massey et al. 2007, 243–249). Furthermore, the trope obscures the socioeconomic diversity among African immigrants (Kusow et al. 2016, 121–129) and it does not attend to how one's African region of origin and US racial hierarchies continue to influence African immigrants' employment and salary attainment for black people in the US (Bideshi and Kposowa 2012, 121–128). Moreover, contrary to popular belief, it is not a zero-sum game where some black people win only because others lose (Lindsay 2015). Ultimately, I conclude that the promise of exceptional status, as a model minority or otherwise, is made in bad faith.

I (Rachel) am aware of the many unearned privileges given to me because of my proximity to middle-class whiteness (my white mother was an academic), my educational privilege, my light skin, and my way of being in the world, which is interpreted within the academy as not-too-disruptive due to existing structures of anti-blackness. Although I have long embraced a political identity as Black, and a politic informed by black feminist ideology and black literature as theory, because I am also a first-generation African I find myself occupying the space of the model minority in institutions, where I see an obvious parallel with first-generation South Asian Texans.

As African feminist scholars, we recognize that the social and political constructions of black identity have placed people in grave peril in the US. Each time we see a new headline about police violence, we are reminded of how the devaluing of black life puts us, our family, and friends at risk. As black women in Texas, we are aware of the precariousness of our own physical safety due to the state's legalization of the carrying of firearms on our college campus.[5] We recognize that we are unable to escape what Patricia Hill Collins (1990) effectively referred to as "controlling images" of Black women, which we must navigate no matter how good a professor or student we are able to be. We identify ourselves as black and as diasporic Africans. We recognize that our identities are created by our positions within national and international contexts and within the social, cultural, and economic contexts that construct race. And we find ourselves simultaneously in danger and uniquely privileged as African women in North American institutions, seen as possessing a more acceptable blackness in the academy–that offers diversity but remains pitted against African Americans (Waters, Kasinitz, and Asad 2014). But we reject the simplified construction of Black/black identity that would pit us against other black people in these hostile institutions. We also reject the anti-black racism in our own African communities, which simultaneously embraces African American culture on the continent and in the diaspora but disparages African Americans.

Eventually we realize that each of us is treated in academia as a "model minority" for our level of education as well as our assimilations to white culture—we are well-versed in the dominant language of academia yet provide an appealing difference. Our multicultural identities serve many purposes in academia (Busia 2018). *We efficiently diversify the space.*

Some Assumptions about Mobility for African Women in the Academy

In the voice of a black, female PhD candidate who enthusiastically began a doctoral program and, later, struggled to secure funding to complete it.

As an African woman in the US academy, I am constantly aware of the hierarchy of theories in the English department. Despite the number of faculty "of color," who are mostly of Latinx, East Asian, and South Asian descent, the English department I am affiliated with, like many others, continues to operate with white supremacist rhetorics and structures. Emails about "diversity" and "equality" and "engagement" circulated in the department long before my arrival, yet the department is resistant to change and hostile to criticism. One incident serves as a good example. During my first three years in the program, all graduate students with teaching fellowships had to attend a pedagogy training. Each year, staff from the Language and Culture Center (LCC), a separate department, explained that we should not be surprised if students "from other countries do not understand critical thinking" and citation practices. One staff member said, "they don't really have that over there." In the third year of this presentation, a peer from Zimbabwe broke into a loud stream of protest, condemned the LCC for being racist and elitist, and then stormed out. Her words were followed by those of a peer from Lebanon—the daughter of Palestinian refugees—and she too stormed out. That year, the staff member who gave this presentation was the lone Black man at the LCC. I had observed these presentations for years and he did not alter any of the content from previous years. It didn't occur to me until later that he may have struggled with the task of presenting racist caricatures of the student population he served, and that faced with that struggle, he accepted it as an "uncomfortable" but "manageable" part of his role at the university. But there was no evidence of such discomfort or struggle at that time. Standing on either side of him were two, middle-aged, white women who had given the presentation in preceding years. Faced with the direct criticism of my peers, he tried to deflect with "good humor." Since I remained in the auditorium, I explained again why his words were a problem, if not an anathema, to the entire English department and our mission as instructors. Later, during a coffee break, a Department Fellow asked me if I thought what I had "done" was okay. She said, "I think you embarrassed him, you know" and went on to suggest that I apologize. I declined to do so as a small group of graduate students lingered around us. She responded by rephrasing her explanation of what I had done wrong. Eventually, I walked away. The department canceled this part of the "pedagogy training" after that year, but other issues with the department, like teaching assignments and required courses for degree plans, linger.

My experience as a graduate student has made me aware that the disregard for difference and historical specificity as well as the easy reparation to ethnocentric explanations that position "foreign" students as less capable is part of what informs the conversations of African feminism. These attitudes about who is capable of what seem to validate ignorance about African feminism. Add to this the way that "disciplinary standards" require students to demonstrate knowledge of major theories and theorists of Western Europe and North America while largely ignoring the work of scholars from Africa, the Caribbean, South America, and, to a lesser extent, India. Though the requirement to include theorists is sometimes presented apologetically, the requirement remains: one must know these white men and what they say in order to be a legitimate scholar and thinker. As I worked to meet the requirements of the discipline, which includes long exams, oral defenses, and a dissertation, I found myself struggling to balance disciplinary expectations with my own way of seeing, understanding, and communicating. The requirement that one must master the discipline and demonstrate the mastery in writing threatens the very subjectivity of the scholar-in-training (Nzegwu 2003, 103–147; Christian 1988). The question asked of those of us who would like to refer to an uncited archive may be, "Can you really do the work of literary and cultural criticism if you cannot read and refer to the canonical scholars?" Through the prolonged "intellectual labor" of dissertation writing, I am more aware that I do the work of African feminist methodology and theorizing *with and against* the discipline of US literary studies. I continually bring African feminist thought into the forefront of my teaching and publications and introduce faculty and colleagues to it so that it will not be overlooked or dismissed by people who otherwise refuse to engage with it. I must create my own legibility in the academy. *This becomes part of my labor*.

I am also a person who "diversifies" the space. I am the instructor that students don't want to have *or* the one they will complain about; the one who dares confront them with uncomfortable truths; the person who interrupts the presentation with my b(l)ack talk. I am also the instructor who must simultaneously teach classes at local community colleges because my teaching fellowship doesn't cover the cost of living. Yet, I also acknowledge the advantages that the US academy affords to me. Because I am in the US, I have access to lines of publishing, to national and international conferences, to a well-organized and digitized library network, and a myriad of other infrastructures, which facilitate intellectual labor. I am also the person who occupies a position of structural privilege in comparison to (some) emerging scholars on the continent. Thus, I acknowledge that I benefit from structures and opportunities due to my position in the US, as I acknowledge my experience in a black body and with a dissenting (black) voice. However, like so many emerging academics, I am mistrustful of the allure of upward mobility and institutional praise. Even before the COVID-19 crisis led to hiring freezes and the elimination of job postings, I was aware that my ability to access the privileges of the US academy would remain fraught. My observation that the breadth of black women's feminisms is not acknowledged and the theories are underutilized is one reminder of this predicament. Despite this awareness, I am encouraged by African feminist theory and praxis to think about how my work might connect with that of others. Specifically, I think about an African feminist praxis that calls on the theorist-scholar-creative to engage with others, to draw from their ideas ethically, to bring their ideas to the fore with care, to engage in meaningful criticism, and to develop better insights from the criticism offered to them. Dr. Quinn and I have done this at each stage of our collaboration.

In the voice of a black faculty member whose graduate school funding options also dissipated in the final year of her studies and for whom it was the few Black women professors on her campus who stepped in to make sure she was able to complete her doctoral degree.

My efforts to build relationships of trust and co-learning beyond my institution have been far more significant to my intellectual growth as a black feminist scholar than those at my institution. Beyond my university, I am released from the institutional pressures that inform my interactions with colleagues, and I am able to build feminist collaborations nationally at the National Women's Studies Association conference, as part of the Future of Minority Studies network, and as a participant at the Thinking Transnational Feminism summer institute and locally among youth and community organizers in SAYHU. As I have directed my attention to race and gender in the US and the Spanish-speaking Caribbean, I have focused less on African feminist thought in my writing and research. However, I have invariably taught about it in global studies courses given the presence of contemporary African feminist thought in popular transnational media discourse. This is accessible to me via African feminist WhatsApp, Instagram, and Twitter networks, where content shared widely includes the emotional justice theory by British-Ghanaian Brooklynite Esther Armah, speeches and strategies by the late Kenyan environmentalist Wangari Mathai, anti-colorism messages from Yaba Blay, and sex-positivity from Nana Darkoa Sekyiamah and Malaka Grant.

Some Questionable Assumptions about African Feminist Theorizing

In the voice of a black, female PhD candidate trained in literary and critical theory.

As I use African feminism as a theory and methodology, I encounter several pervasive assumptions in the US academy. The assumptions may be expressed in different ways but they fall into three big groupings. The first assumption is that the theoretical scope and relevance of African feminism is limited to Africa. The second assumption is that it is a new or underdeveloped theory, and the third assumption is that African feminists are few and far between. These assumptions are reproduced in graduate seminars, in "informal" discussions at department events, in responses to funding requests, and so on. Their effect is to put African feminisms into question. The latent question seems to be: "Does such a feminism exist? If so, to whom is it important or impactful?"

The answer, of course, is that African feminisms do exist. They exist in the plural and cannot be reduced to a singular theory. African feminisms are transnational and reflect the globalized flow of people, ideas, capital/currency, goods, and resources. African feminisms are responsive to the economic and political conditions of women, children, and minoritized subjects.[6] Moreover, African feminists are not few and far between. Lastly, all of the individuals whose behaviors, ideals, and actions strive toward gender equity do not call themselves "feminist" due to various reasons (Atanga 2013). Despite these differences, the theoretical underpinning of African feminism is rigorous.

In addition, over the last two decades African feminist theorizing has moved to online spaces where it is possible to reach a broader audience. The work of Nkiru Nzegwu, which spans more than two decades, provides an apt example. Nzegwu is a philosopher and scholar of art, Africana Studies, and African feminisms. Nzegwu founded Africa Resource Center in 1999 without any institutional funding (Nzegwu 2002, 81). The initiative went on to launch five academic, peer-reviewed journals, including *JENdA: A Journal of Culture and African Women's Studies*. It also launched AfricaResource.com, a public-facing website, which presents articles on a wide-range of subjects, including art, culture, literature, gender, race, current affairs, and science and technology. As Nzegwu argues,

> content is the area with the greatest need in the near future. We need to think simultaneously of the kind of information that will be on the Internet. If we are not part of generating that knowledge, we cannot expect it to reflect who we are in a positive way.
>
> (82)

Nzegwu's African Resource Center (ARC) speaks both to the possibilities of ongoing initiatives and the possible challenges of a digital and public-facing project that centers Africa, gender, feminism, and culture. For example, AfricaResource.com has articles on a wide breadth of topics and covers a period of over a decade, but the infrequent publication of new material illustrates the difficulty of funding and maintaining a public-facing website. As Nzegwu has noted, access to funding is not universal. As institutions recodify assumptions about who is and who is "not Internet savvy, imaginative, and resourceful" (2002, 81), they exclude those who might otherwise contribute to public humanities and digital feminisms. Nzegwu, "an over-the-hill immigrant African woman of the wrong color with an unpronounceable name, doctoral degree in the humanities, and teaching position in some lackluster public university" (81), understands herself to be one of the many excluded by the "logics" of ethnocentrism, misogyny, and racism. By reviewing Nzegwu's creation of ARC and its challenges, we gain a more accurate view of the challenges facing African feminists who entered public scholarship in the digital form during the late 1990s.

Joined here by the voice of an all-but-tenured faculty member at a research institution, who has sustained an ambivalence about academia that has led her to invest in public-facing and community-based projects through digital feminist networks, where her intellectual contributions are more legible.

Working with Maurine while she is immersed in African feminist theory for her dissertation has been an opportunity for me to recalibrate my own intellectual focus. Connecting with other African feminists in academia has allowed me to finally see myself as an African feminist in my own right, in conversation with African feminists on the continent and in the diaspora. From this positionality, I have also pushed Maurine for a queer reading of contemporary African life throughout the literary texts that she has selected to study. This is especially necessary with an author like Adichie, who has come to represent a globalized and consumable contemporary African feminism yet is trans-exclusionary. I teach Adichie's work with the caveat that I question her transphobia head-on and work against a circumscribed notion of African feminism today (Camminga 2020, 233).

Nzegwu's ARC described above contrasts with more recent online publications that bring feminist work to broad audiences across the globe—some via dedicated platforms and others not. More platforms are now focused on publishing only on African feminist thought, and some of these are supported by or benefit from global development networks and infrastructure. For example, writer and co-founder (with Malaka Grant) of the blog *Adventures from the Bedrooms of African Women*, Ghanaian feminist Nana Darkoa Sekyiamah is also the Communications Specialist for the Association for Women's Rights in Development (AWID) and previously Director of Communications and Media for the African Women's Development Fund (AWDF). Likewise, *The African Feminisms Blog*, initially founded by Billene Seyoum Woldeyes as *ethiopianfeminisms*, includes articles, reviews, and criticism. Seyoum, who is today the Foreign Press Secretary for the Office of the Prime Minister of Ethiopia, resides on the African continent, a location from which she has increased capacity for digital networks of feminist theorists and gender studies scholars across the continent and in the diaspora. Most of the African feminist scholars we point to live in the US or UK but like us, keep connected to contemporary African feminist knowledges through digital tools now available to them.

Twenty years into the millennium, African feminist thought online has benefited from the accessibility and packaging of sleek, seemingly ubiquitous websites that are low-cost or no cost to users for self-publishing, for example, Awino Okech's reflections on Medium.com. Okech's pithy entries offer interventions on gender, racialization, institutional racism, mentorship, and so on. She also posts reading lists that feature gender and feminist scholars from the continent and links to open-access texts.[7] Meanwhile, *Black Women Radicals*, founded by Jaimee A. Swift, has a reading list of African feminisms and an online conversation between Nana Afua Yeboaa Brantuo, Terza A. Silva-Neves, and Namupa Shivute, which provide an invaluable entry point into the voices of many transnational black feminists ("African Feminist Perspectives Matter"). Institutional partnerships negotiated by African scholars within the North American academy allow for greater support through online publication of scholars from across the continent, as is the case with South African scholar Sean Jacobs' website *Africa is a Country*. The African Feminism series that Rama Salla Dieng recently curated on the site, with a literature review of divergent African feminisms (with links), is invaluable for teaching the field and making the discipline visible. At the same time, as a development studies scholar based in the UK, Dieng's broader effort to engage a contemporary audience asks, "How are young feminist scholars using their life experiences as sources and resources for theorizing their feminism?" Our examples above provide only a partial representation of the current breadth of African feminisms online, while other conversations circulate via social media networks, text threads as well as recorded roundtables captured on Zoom and sometimes shared in the digital archive of YouTube.

Ways of Thinking about Digital Humanities and Public Humanities

In the voice of an overextended and untenured black faculty member who teaches black studies from a gender studies program and is interested in challenging anti-blackness beyond the academy because institutional will is limited.

My research and teaching have focused on transnational feminisms, visual culture, and sexualities in the African diaspora, in particular Dominican women's transnational identities and experiences of race and gender in Santo Domingo. I bring to the digital humanities project and working group Black.Migration. Houston. (B.M.H.), a black feminist methodology and curiosity about community-building across digital networks as well as a desire to engage with black feminist knowledges to develop our work.[8]

My inclination as a scholar-activist has been to create spaces for engagement and cross-pollination among community members and my colleagues; over the years, I have coordinated interdisciplinary working groups on Caribbean film, queer studies, and *afro-identidades* for collaborative learning among scholars. With B.M.H., I have expanded my work in the digital humanities and by coordinating students, community members, and colleagues. First imagined as a collaboration with the national organization Black LGBTQ+ Migrant Project (BLMP), inspired by a conversation with Houston-based Nigerian activist Ola Osaze, B.M.H.'s objective has been to create a digital resource for black LGBTQIA+ migrants and those seeking to provide them services, particularly in Houston and in the Texas Gulf Coast region. Through my connection with BLMP, I have understood how current national attention to migrants at the Texas-Mexico Border does not account for how anti-blackness and homophobia inform discriminatory policies worldwide. Yet Africans and Haitians represent a significant percentage of migrants in detention facilities, a large number of which are located in Texas (Ahmed et al., 2018). At the same time, the Transgender Law Center notes that homosexuality and gender non-conformity or sexual difference is criminalized in 38 out of 54 African countries, contributing to forced migration due to homophobic/transphobic violence, economic precarity, lack of legal status for citizenship, environmental racism and more (Transgender Law Center). Asylum seekers in the US then face additional anti-blackness, homophobia, and transphobia as they attempt to navigate US institutions while seeking to live free of state violence (Transgender Law Center). In an effort to respond to uprisings against police violence and anti-blackness around the country in 2020, I began to explore a growing number of funding opportunities that could support a digital humanities pedagogies and critical race theory project such as B.M.H.[9] Our website is now a hub from which to circulate a national survey for BLMP. It will eventually link an increasing number of local services and tools to assist black queer migrants in navigating migration, asylum policy, and legal systems as policy shifts. The website will eventually host an interactive reading room around each of the topics we explore and direct visitors to resources, websites, and course curriculum highlighting issues of migration, blackness, and sexuality.[10]

Building on the notion that "digital humanities offers students new approaches to the multiculturalism, multimodalities and multimedia," DH scholar Roopika Risam argues that her concept of "postcolonial digital pedagogy helps students develop an emancipatory digital cultural literacy—and awareness of how digital knowledge production is imbricated in the politics, power and neocolonial practices that privilege the epistemologies of the Global North" (2019). Like many DH researchers and scholars, my approach is one of collaborative learning and experimentation. I convened an interdisciplinary group for the purpose of cross-pollination, including experts in new media from our neighboring community college. I brought new people into the conversation, offering peer mentorship and support for new projects that center migration, blackness, and sexuality. A DH project like ours serves as a form of collective publication of feminist knowledge and is decidedly non-linear in its formation.[11] I see the impact of the project as multilayered: networks and resources for black LGBTQIA+ migrants, more attention to these issues for migrants among colleagues and the general public, knowledge-sharing, as well as training and mentorship for Maurine in DH and public scholarship in the academy. To this end, we located a small amount of funding for graduate student labor that allowed Maurine to swap out teaching responsibilities for project development and project management and apply her website construction skills to BMH.

Our working relationship is different from what it might have been had I only served on her committee. In working with Maurine on this project, I have pushed her to reexamine her research and role as an educator beyond what she was trained in and to expand her skillset beyond her formal training in literary theory. As we strategized the project, we strengthened our collaboration and each adapted to working with one another, work that takes time and attention. Over the course of a year, we have been able to move into a collaboration that allows me to impart knowledge about public scholarship and apply and model a black feminist leadership methodology. I invest my time in her success in part because it aligns with my core values around applied theory. Each time I do so, I recall the innumerable and often discrete moments when many other scholars of color stepped in to help me survive and succeed in the academy.

> *In the voice of a black female PhD candidate who is relatively new to the digital humanities.*

When I came to the B.M.H. working group through a research assistant opportunity, I had limited experience with public-facing scholarship because I was trained in literary and composition pedagogy and theory. My initial months as a research assistant were focused on learning more about the topic, the goals of the collaboration, and the co-working aspect of the assistant role. The multiple foci meant that I was less focused on process and the connection to DH theory. I have experienced both the challenge of thinking about differences in working and communication styles as well as the challenge of thinking through digital humanities. Ultimately, I follow Dr. Quinn in seeing African feminist thought as a theory of collaborative labor. The details of how the theory becomes action may change from one person to another, but it is a useful model for ethical collaboration.

One transformative aspect of the work is the emphasis on continual development. This is a move away from the "complete by date" model that I have used successfully to meet benchmarks throughout graduate school. The "complete by" model emphasizes mastery of a concept and the conclusion of a project. Instead, Dr. Quinn and I have been intentional about the ethic of ongoing development; our digital humanities project is feasible because it allows for continual revision. Each member can contribute "outside of their disciplinary expertise." Each member can make changes (including additions and deletions) to the written or visual content, which requires a trust-as-action non-hierarchical approach. I recognize this as a strategy for surviving within the academy. By creating equality among collaborators, including non-academics and artists, the possibility for mutual and multiple investments in the project becomes possible.

Some Reflections on Doing Public Scholarship Together

> *In the voice of an all-but-tenured faculty member at a research institution* and *the voice of an underfunded black PhD candidate committed to producing creative and critical texts. In the voice of two African feminist scholars who recognize that we have power in numbers.*

As other African feminist scholars have demonstrated, it is critical for us to reflect upon how we do the work. This essay offers us another layer of collaboration and the opportunity to think together as African feminist scholars doing public scholarship through a DH project in the South. It complements a black feminist pedagogical praxis that each of us is committed to as scholars, no matter where we are in our academic trajectories. Misogynoir within the academy reminds us to look for other black women able to speak to our experiences and stand with us while our intellectual commitments to African feminism and African literature connect us across other differences. Because of our disciplinary backgrounds and particular working styles, we have kept an open dialogue around the ethics that we draw upon to do public-facing work with scholars and community members. Like the structure of this essay, the work within and beyond BMH is a feminist experiment. Ultimately digital publication and digital humanities allow us to build networks with feminism scholars and activists on the continent and in the diaspora, with a radical politics of citation and resource sharing.

I (Rachel) have made communication a priority as a mentor and collaborator. By taking time away from my own publication deadlines, I invest in students who might carry forward our shared values. I proposed that Maurine and I attempt to write this joint essay in order to reflect on our collaborative work with B.M.H. and solidify the methodology we have discussed. Ultimately, I understand personal and professional reflection on process as a part of our metrics of success. My approach to mentorship is firmly wedded to notions of support and equity that are often dismissed in the academy in lieu of false urgency, scarcity of time, and a deliverables-oriented approach. When normalized as an aspect of the mentoring relationship, my personal investment in graduate students of color recognizes them as in need of comprehensive guidance on how to navigate white supremacist heteropatriarchal institutions.

Public scholarship makes it possible for us to imagine an alternative future for black women as intellectual laborers. It also provides the means to create an alternative life or archive for the African feminists who have preceded us. And it requires that we talk b(l)ack. I (Maurine) have drawn upon my understanding of African feminist labor (and expanded it) as I have worked with the scholars, visual artists, and community organizers who make up our working group. Dr. Quinn repeatedly acknowledges the collaborative members as engaged, necessary, and uniquely positioned to contribute to the study of black and LGBTQ+ migrants, and this has set the tone of our meetings. As a graduate student I am recognized in this academic space not only as someone who contributes labor but who has her own ideas to share and her own scholarly ambitions. As we recreate African feminist theory together through transformative action, we help a critical transnational black feminist pedagogical praxis take shape.

Notes

1 I, Rachel, I use "Black" as it is capitalized in the US today, to mark the *political identity of African Americans*, however in writing about "black" identity across the African diaspora I use the lowercase "b" of Stuart Hall's cultural studies *diasporic black identity*.
2 I theorize this identity in relation to Ghanaian feminist scholar Abena Busia in my 2021 essay "Afi is for Us" for the lesbian feminist journal *Sinister Wisdom*.
3 The Arts of Citizenship program at University of Michigan influenced my graduate study, and later, the related national organization Imagining America (IA) provided my first exposure to the language of public scholarship. IA's work includes an effort to shift the landscape of academia and teach university administrators to value community-based projects when evaluating scholarly contributions.

4 As I have discussed at the National Women's Studies Association conference, many of my South Asian mentors and peers have found feminist co-conspirators among black feminists: Chandra Mohanty with Linda Carty, Beverly Guy-Sheftall, Jacqui Alexander, and others; Shreerehkha Subramanian with Abena Busia and Demita Frazier, for example.

5 Senate Bill 11, also known as the "campus carry" law, codified in Texas Government Code, Section 411.2031 the right of licensed gun owners to carry concealed handguns on public college and university campuses. Though the law claims that universities "may enact reasonable rules and regulations," as an instructor Maurine was forewarned that banning guns in her class would be a violation of Texas State Law.

6 I, Maurine, follow scholars such as Filomena Chioma Steady who understand feminism not only as a theory but also, importantly, as an aspect of women's lives, behaviors, actions, and choices. Writing about feminism and women's associations and collective action groups in *Women and Collective Action in Africa*, Steady argues that "Given the legacy of colonialism and the current destructive trends in corporate globalization, feminist intentions are intricately bound up with the practical exigencies of economic domination" (2005, 6). Thus, my use of feminism refers to the economic base of society and women's actions. See also *Development Dialogue* for articles by feminists, including Steady, which advance this view for countries of the Global South.

7 See, for instance, Okech's "A Feminist Reading List on Care, Crisis and Pandemics" (2020).

8 I draw on the wisdom of Combahee River Collective co-founder Demita Frazier, interviewed at University of Houston-Clear Lake by Shreerekha Subramanian in March 2020 at the beginning of the global coronavirus pandemic.

9 Although the 2020 uprisings against police violence towards black people spawned "Grants to Enhance Research on Racism" at our university with hundreds of thousands of dollars disbursed, our project was not funded.

10 Early B.M.H. collaborator Soledad Álvarez Velasco spearheads a project on migration with a slew of other scholars from across the Americas—a region that African migrants cross to arrive in the US; however, few examine anti-blackness in their research or engage African feminist theory. See Álvarez Velasco and Lucía Pérez Martínez, "Intervention," 2020.

11 The non-linearity of digital projects resists being forced into linear form—African scholars experience a similar force from hegemonic whiteness. Nzegwu (2003) notes that sequence is not always the most effective way.

References

African Feminism. 2018. Accessed 1 February 2021. http://africanfeminism.com/.

Ahmed, Manan., Maira E. Álvarez, Sylvia A. Fernández, Alex Gil, Merisa Martinez, Moacir P. de Sá Pereira, Linda Rodriguez, and Roopika Risam. 2018. "Torn Apart/ Separados." Accessed 1 February 2021. http://xpmethod.columbia.edu/torn-apart/volume/2/index.

Álvarez, Velasco Soledad, and Lucía Pérez Martínez. 2020. "Intervention – 'Pandemic and (Im)Mobility in the Americas'." *Antipode Online*. Accessed 1 February 2021. https://antipodeonline.org/2020/08/11/pandemic-and-immobility-in-the-americas/.

Anderson, Monica. 2015. "6 Key Findings about Black Immigration to the U.S." *Pew Research Center*. Accessed February 11, 2021. https://www.pewresearch.org/fact-tank/2015/04/09/6-key-findings-about-black-immigration/.

Armah, Esther. 2020. "What is Emotional Justice." *The Armah Institute of Emotional Justice*. Accessed February 1, 2021. https://www.theaiej.com/emotional-justice

Atanga, Lilian Lem. 2013. "African feminism?" In *Gender and Language in Sub-Saharan Africa: Tradition, Struggle and Change*, eds. Lilian Lem Atanga, Sibonile Edith Ellece, Lia Litosseliti, and Jane Sunderland, Jane, 1–26. Amsterdam: John Benjamins Publishing Company.

Bideshi, Davison, and Augustine J. Kposowa. 2012. "African Immigrants and Capital Conversion in the U.S. Labor Market: Comparisons by Race and National Origin." *Western Journal of Black Studies*, 36(3): 181–200.

Brantuo, Nana Afua Yeboaa, Terza A. Silva-Neves, and Namupa Shivute. 2021. "African Feminist Perspectives Matter: A Reading List." *Black Women Radicals*. Accessed February 3, 2021. https://www.blackwomenradicals.com/blog-feed/african-feminist-perspectives-matter-a-reading-list.

Busia, Abena P.A. 2018. "Creating the Archive of African Women's Writing: Reflecting on Feminism, Epistemology, and the Women Writing Africa Project." *Meridians*, 17(2): 233–45.

———. 2003 "In Search of Chains Without Iron: On Sisterhood, History and the Politics of Location." In *African Women and Feminism: Reflecting On the Politics of Sisterhood*, ed. Oyèrónké Oyewùmí, 257–68. Trenton, NJ: Africa World Press.

Camminga, B. 2020. "Disregard and Danger: Chimamanda Ngozi Adichie and the Voices of Trans (and Cis) African Feminists." *The Sociological Review*, 68(4): 817–33.

Christian, Barbara. 1988. "The Race for Theory." *Feminist Studies* 14(1): 67–79.

Collins, Patricia Hill. 1990. *Black Feminist Thought: Knowledge, Consciousness, and the Politics of Empowerment*. Boston: Unwin Hyman.

Development Dialogue. 1982. Accessed June 10, 2023. https://www.daghammarskjold.se/development-dialogue/

Dieng, Rama Salla. 2020. "Talking back: African feminisms in dialogue." *Africa Is a Country*. Accessed February 11, 2021. https://africasacountry.com/2020/05/talking-back-african-feminisms-in-dialogue.

Kusow, Abdi M., Sitawa R. Kimuna, and Mamadi Corra 2016. "Socioeconomic Diversity Among African Immigrants in the United States: An Intra-African Immigrant Comparison." *Journal of International Migration and Integration*, 17: 115–30.

Lindsay, Keisha. 2015. "Beyond 'Model Minority,' 'Superwoman,' and 'Endangered Species': Theorizing Intersectional Coalitions among Black Immigrants, African American Women, and African American Men." *Journal of African American Studies*, 19(1): 18–35.

Lugones, Maria C. and Elizabeth V. Spelman. 1983. "Have We Got a Theory for You! Feminist Theory, Cultural Imperialism and the Demand for 'the Woman's Voice'." *Women's Studies International Forum*, 6(6): 573–81.

Massey, Douglas S., Margarita Mooney, Kimberly C. Torres, and Camille Z. Charles. 2007. "Black Immigrants and Black Natives Attending Selective Colleges and Universities in the United States." *American Journal of Education*, 113(2): 243–71.

Mitchell, Koritha. 2018. "Identifying white mediocrity and know-your-place aggression: a form of self-care." *African American Review*, 51(4): 253–62.

Nnaemeka, Obioma. 2003. "Nego-Feminism: Theorizing, Practicing, and Pruning Africa's Way." *Signs: Journal of Women in Culture and Society*, 29(2): 357–85.

Nzegwu, Nkiru. 2002. "Africaresource.com: Bridging the Digital Divide." *African Issues: The African "Brain Drain" to the North: Pitfalls and Possibilities*, 30(1): 81–5.

———. 2003. "O Africa! Gender Imperialism in Academia." In *African Women and Feminism: Reflecting on the Politics of Sisterhood*, ed. Oyèrónkẹ́ Oyěwùmí, 99–157. Trenton, NJ: Africa World Press.

Okech, Awino. 2020. "A Feminist Reading List on Care, Crisis and Pandemics." *Medium*. Accessed February 11, 2021. https://awino-okech.medium.com/a-feminist-reading-list-on-care-crisis-and-pandemics-eaf77a9d293b.

Oyewùmí, Oyèrónké. 2003. "The White Women's Burden: African Women in Western Feminist Discourse." In *African Women and Feminism: Reflecting On the Politics of Sisterhood*, ed. Oyèrónké Oyewùmí, 25–44. Trenton, NJ: Africa World Press.

Quinn, Rachel Afi. 2022. "Afi is for Us!" *Sinister Wisdom: A Tribute to Conditions*, eds. Cheryl Clarke, Julie Enszer, Shromona Mandal, 123: 202–8.

Risam, Roopika. 2019. *New Digital Worlds: Postcolonial Digital Humanities in Theory, Praxis, and Pedagogy*. Evanston, Illinois: Northwestern University Press.

Rodriguez, Cheryl Rene, Dzodzi Tsikata, and Akosua Adomako Ampofo. 2016. "Introduction." In *Transatlantic Feminisms: Women and Gender Studies in Africa and the Diaspora*, eds. Cheryl Rene Rodriguez, Dzodzi Tsikata, and Akosua Adomako Ampofo, vii–xxxii. Lanham: Lexington Books.

Senate Bill 11 (SB 11), § 411.2031: Carrying of Handguns By License Holders on Certain Campuses, Tx. Government Code (2015).

Steady, Filomena Chioma. 1981. *The Black Woman Cross-Culturally*. Cambridge, MA: Schenkman Pub. Co..

———. 2007. "The Black Woman and the Essentializing Imperative: Implications for Theory and Praxis in the 21st Century." *Race, Gender & Class*, 14 (1/2)): 178–95.

———. 2005. *Women and Collective Action in Africa: Development, Democratization, and Empowerment, with Special Focus on Sierra Leone*. New York: Palgrave Macmillan.

Ukpokodu, Omiunota N. 2018. "African Immigrants, the 'New Model Minority': Examining the Reality in U.S. K-12 Schools." *The Urban Review*, 50(1): 69–96.

Waters, Mary C., Philip Kasinitz, and Asad L. Asad. 2014. "Immigrants and African Americans." *Annual Review of Sociology*, 40: 369–90.

Section II

Black Women's Lived Experiences in Our Contemporary Societies

4 Mothering in Neo-Liberal Contexts

Caribbean Women's Experiences

Daniele Bobb

Introduction

The issue of women's work in the Caribbean has been explored by many Caribbean feminist scholars (Dunn 2004; Reddock 1994; Trotz and Peake 2001). Most point to patriarchal ideology as underpinning the categorization of women and the specific devaluation of social reproduction (Trotz and Peake 2001). Leith Dunn (2004), in particular, in her examination of the policy implications of the International Labour Organization convention for employed women, highlights the peculiarities of women and how they are used as tools of discrimination. Mondesire and Dunn, in their report to CARICOM Secretariat on the status of Caribbean women in 1995, argue that "women's unwaged and often invisible work in the home and in communities is another source of power which has not been fully tapped: by quantifying their voluntary labor, women stand to gain social and economic recognition for their efforts" (Mondesire and Dunn 1995, 81).

These scholars, I argue, begin from the flawed premise of seeking to address the inequality and injustice of women by inserting women's domestic activities into patriarchal (and capitalist) valuing of work, with the belief that it will be empowering for women. Their focus has not been to question the conceptualization or structure of work, that is, historicizing and analyzing how it is understood, but rather, they argue for the valuing of household "work", that is, the monetary value and understanding of work.

In historicizing work, Keith Grint (1991) explores the various manifestations of the socio-cultural construct, from a necessary and imposed activity to one that is motivated by possible gains, both earthly and heavenly (protestant ethic). Initially presented as something to be avoided, an activity only for the enslaved, work eventually became expressed as an activity through which one's goals and desires can be attained, as through earning money, one can purchase items that are markers of the "good life" (Grint 1991; Weber 1992). Engaging in paid work became a norm. Still, work done in the home continues to be devalued, invisible and taken for granted, as work was defined as a public activity that has an economic value. Since the 1960s, feminists have been troubling this conception of work as public as it excluded women's care and domestic

DOI: 10.4324/9781003143550-7

activities and reproduced gendered inequalities such as the gender pay gap, and women's exclusion from certain professions perceived as "male jobs" (Nash 1995, Firestone 1997; Holmstrom 2003).

Two decades into the twenty-first century women's work continue to be on the feminist agenda for achieving gender equality. Kathi Weeks believes this is because of feminists' uncritical adoption of the concept "work" without unpacking it (Weeks 2011). Examining work from a political perspective, she historicizes the concept, specifically the values that inform the formation of the concept "work." She, therefore, argues that the concept shapes the way work is imposed on the body, how it is enacted, and how it is perceived and valued (Weeks 2011). Weeks regards work as an activity of production which sometimes can be leisurely and which is more than just economic but is also social, cultural, historic and political, that is, she sees "work" as dense.

Using Barbados as a case study/example, this chapter illustrates the multiple imbrications of work as well as its relationship with mothering and motherhood in the Caribbean.[1] This is done through an exploration of select in-depth interviews conducted (in 2014 with follow-ups in 2016) with 21 (Caribbean Community) CARICOM women residing in Barbados. Herein, work is not presented as oppositional to mothering, or in a binary manner, but rather as multilayered with multiple uses, producing a worker subject that co-exists with the maternal subject. It examines the confluence and divergence of the maternal and worker subjectivities as they exist coterminously. In illustrating the reciprocal relation of economic, gendered and classed relations of power, I identify points at which maternal and worker subjectivities become compatible yet digress. I capture how women's normalized perception of "work" influences their mothering, and their relationship with work. Thus, I draw on the multiple ways in which skills associated with mothering are called upon by the worker and employer, advancing the interest of their workplaces. I then compare how certain employment skills are utilized within maternal practices to both challenge normative maternal practices and reinscribe hegemonic maternal discourses.

Regulating Maternal and Worker Subjectivities

Barbados is a small Eastern Caribbean Island of 167 square miles and is situated in the Northern Atlantic Ocean at 13°10′ north of the equator. In 2020, this English-speaking Commonwealth Island had an estimated population of 280,693 people (World Bank Data 2023). Its main income-generating industry is tourism. Barbados, a former British colony, gained independence in 1966 and in 2021 became a Republic. It has 92.4% Black, and Barbadians are ethnically referred to as Afro-Barbadians as they are descendants of enslaved Africans (Barbados Integrated Government 2023).

Like many capitalist societies, Barbados' economy is grounded in neo-liberalism, an ideology that advances competition and individuality as the defining characteristics of market and human relations. The 1995–2001 period in

Barbados saw improvements in living conditions, with steady but moderate economic growth and a falling rate of unemployment. Barbados had a strong social protection system which has been a defining feature of Barbadian policies. Like the international recession of 1980–1982, the 2007 economic recession severely reduced the demand for several of the region's main exports and lowered the number of tourists visiting the island. Thus far, the beginning of the twenty-first century has seen a balance of payment crisis in Barbados and great debt which has resulted in the closing of many private businesses and the lay-off of many people. As the government prepared to access a loan from the International Monetary Fund (IMF) in 2018, it made certain structural adjustments within the state. This was seen by the World Bank as "varied policy action that attempts to alter the nature, structure and functioning of economies," (Antrobus 2004, 68). The policies are based on neo-liberal economics that saw an expansion of private sector influence over policy and development and are concerned with restoring macroeconomic stability with a view to promoting economic growth. With structural adjustment programs (SAPs), it was expected that strong economic growth would benefit the poor via a trickle-down effect. However, this translated into allocation cuts by the government to important sectors such as education, health welfare and agriculture, cut in subsidies to the poor, and privatization of public services and public sector investments (Antrobus 2004, 68). These changes removed many state assistance programs and reproduced unequal gender relations of power by positioning mothers as the sole provider for their children. These economic, cultural and social components of neo-liberalism ensure mothers' navigation and negotiation of motherhood are not just difficult but also complex in that the neo-liberal ideology of motherhood constructs mothers as "both responsible for their families and incapable of that responsibility" (Apple 1995, 162).

The complexity of motherhood is demonstrated in Neela's experience.[2] Neela is a self-identified working-class, primary-level educated, Guyanese mother of two children. As a sewing machine operator in a clothing factory, her monthly income is under USD $500.00. She supplements this by working as a domestic cleaner during her spare time. Unsatisfied by her employment situation, she describes it as a "means to an end" relationship between the maternal and worker subject when she states, "the job don't pay good and they treat you like dog, but I have mouths to feed so I put up with it." Neela believes that the lack of education beyond the primary level defined her economic and social class, as well as limited her options; this is compounded by her immigrant status in Barbados. She accepts "menial work" and inhumane treatment to financially provide for her children. Although this saddened her, she admitted that she would not give up her children for the world and would "take the dog treatment"[3] for her children. Neela's negative emotions are mitigated by her belief in the discourse of good mothering as equating to a sacrificial mother. Affect theorists such as Lauren Berlant (2011) and Sara Ahmed (2010) have written about the attachments one has to certain ideas, values and objects, creating emotional connections that sustain them even when proven to be unachievable (Berlant 2011). In her work on cruel optimism, Berlant describes "cruel optimism" (or the cruelty of life under capitalism) as when a person gets attached to "a cluster of promises regarding the good life that are generated by and embedded in conditions that are already not working" (Berlant 2011, 23). In other words, one becomes attached to conditions that defeat the fulfillment of said promises.

This cruel optimism is evident in Neela's situation, where she is attached to a cluster of promises regarding the good life that are generated by and embedded in conditions that are already not working and reinforce her subjugation not only as a mother but also as an employee. Neela does not challenge the conditions of her employment as she cannot risk losing the job since she needs it so she can take care of her children. Thus, the object that draws her attention (being a good mother by providing financially for her children) actively impedes the aim that brought her to it (her working hours prevent her from spending quality time with her children, leaving her with feelings of guilt). The relations of power produce the worker who complies with the exploitative circumstances of her job to fulfill her maternal expectations and maintain the optimism for a better life for her children. The difficulties she encounters remain uncontested as attachment to her maternal desires remains primary. Motherhood is a dense site of affective, economic and political relations.

The relationship between class, job type and motherhood is also seen in Rachel's experience. Rachel (a tertiary level educated mother who identifies as middle class), who made the decision to not work after her pregnancy, views her absence from the labor force as a privilege. She reflected on her pregnancy stating,

> When I found out I was pregnant I didn't work because I always say with my first child I wanted to stay home at least a year to 18 months with her, to spend the time with her, because I knew, … my mother was working and my grandmother, like before, kept my sister and I because she thought you know, those stages were important. So I just, I guess from hearing that all the time I just thought well let me stay home with her, and it wasn't that rough then but now in this economic crisis you can't do it, but then I could have done it.

This statement highlights two very important points: firstly, it shows how a daughter's experience as a child molded her own values and ideas about parenting her own daughter, and secondly, it demonstrates how changes in the political economy impact the society's culture which in turn affects the choices women *can* make thus suggesting that "choices" are non-linear. Rachel explains that both she and her spouse had "good paying jobs," and they were "financially comfortable." Given the economic climate in Barbados at the time (2007, before the global financial crisis), she felt confident that she would acquire another job after staying at home with her daughter for two years. However, two years later, the economic situation in Barbados worsened, forcing Rachel to accept a job she did not really want, but was necessary for her family's survival. Still, the working conditions Rachel experienced, although not ideal, were much better than Neela's, and as a Barbadian, Rachel did not encounter the immigration issues which forced Neela to remain with her company so she could retain her work permit. Thus, while the expectations of motherhood are homogenous, the realities of mothers are expansive and heterogeneous. "Work" for both women was perceived as a necessary "means to an end" activity, fitting in with normative discourses around the self-sacrificing and financially providing mother, as well as a practical need to provide.

These challenges with income-earning activities are sometimes navigated by women who choose to have children after they have established their careers. For example, Vicky, one respondent in the study, postponed having children and had her first child at age 40. Vicky, at the time of the interview, a 49-year-old mother of two, from Trinidad, recalls that as a young woman she was a "workaholic". She wanted to "help the poor people of the region" and it gave her joy to do so. Propelled by her training, she was determined to positively impact the lives of Caribbean people and she knew it involved a lot of "work" which she believed was impossible if she had a family. Her perception was informed by the gendered conception of work as incompatible with mothering.[4] What Vicky read in books and witnessed in other families shaped what she thought was required of mothers. Again, this illustrates the neo-liberal conception of work and success as transformed by the belief that it requires all of one's time (Giroux 2008). This meant no external distractions (motherhood) as time is perceived as zero-sum, thus positioning the worker identity as dominant and in full control of the capitalist mode of production. Mothering is depicted as a set of tedious and time-consuming activities that are naturally the mothers' and therefore would not permit her to invest the amount of time needed for professional and financial advancement (Apter 1993). This persuaded women to rethink motherhood and to reposition and align themselves with the expected demands of motherhood. Vicky's conceptualization of work demonstrates its complexity as it is both time-consuming and laborious, but also personally rewarding and enjoyable. Marrying later in her life after she acquired adequate material resources and had secured her dream career, Vicky then decided to have children. Vicky's description of this suggests a linear occurrence (job, money, marriage and child). This however is far from the reality for most women and illustrates the cruel optimism of these flawed ideals.

In the interview, Vicky also shared that she experienced complications with both of her pregnancies, and while she did not divulge the nature of her complications, she recalls that her pregnancies were good except for the end, saying,

> The first time around I didn't need the extra (time), because although I had complications, doctors were able to deal with them, I didn't have [a] protracted health situation. The second time around I was on leave for probably more than 4 months because I had complications and then my mummy also died. She knew I was gonna have complications, so she came to visit me, and she died when she was here [Barbados].

Even with complications, Vicky refused additional leave from work as she "didn't want to miss much of work." Her work expectations shaped her maternal practices and choices. It was guided by the belief that: career success requires great time investment in the job (Acker 1990), the gendered structure of the organizations that reward employees for their tenure by using absence from the job as a mitigating matter for promotion (and thus leadership positions), and access to training for advancement (Acker 1990).

Vicky noted that it was during this time that she employed a full-time babysitter as she recognized she was "emotionally drained" from everything that was happening in her life. This recognition and acceptance of her circumstance led her to refute the normalized maternal discourses as she explained, "I knew as a mother that taking care of myself was the best thing for my children, and that is what I did." Yet, Vicky's actions continued to be framed within the maternal script of doing what is "best for the child." The self is only taken care of when the outcome is beneficial for the child. Thus, while her actions appear to be ones of transgression, they are aligned with the institutionalization of motherhood.[5]

Vicky explained that the number and spacing of her children, who are two years apart, was because of the age at which she chose to become a mother, and because she experienced complications which meant longer maternity leave, hence longer time away from her job. Vicky negotiated her career and maternal dreams by compromising her family size. Thus, Vicky's maternal desires are regulated to accommodate her worker subjectivity. By ascribing domestic and care work solely to women, choices and practices of the maternal subject in relation to her career are re/shaped. Thus, while delaying having children/becoming a mother may be perceived as women's "free will" to attend to their personal growth, it is a regulation of their maternity. Even when women negotiate in favor of themselves, it is not transgressive as they often feel unsatisfied, regret many of their choices, and subscribe to the maternal script of good mothering.[6]

Caribbean Women's Negotiations with Work[7]

The historicization of women and work in the Caribbean outlines how Caribbean women were exploited first as (un)gendered dehumanized chattel, as mothers reproducing the enslaved labor force, then by capitalism as cheap labor, and taken for granted thereafter (Bush 2010). Despite changes in the political economy and the international women's rights achievements during the late twentieth century, the progression of women's situation is non-linear. According to Sistren (2005), "where the society appears to move forward, women often do not benefit, and in fact, in the Caribbean it appears as if considerable losses in autonomy are taking place." This comparison by Sistren[8] of women's lives in the pre-independence period in Jamaica and the early twenty-first century is also reflected in my study in Barbados. In this section, I argue that privatized profit-driven economies, although advocating gender equality, engage in actions/activities/practices which are in direct opposition to their advocacy, utilizing a narrative of individual rights and freedoms, encouraging individuals to focus on the self as a means of demonstrating their commitment to equality.

Despite the international treaties from the International Labour Organization (ILO) and the Convention on the Elimination of all Forms of Discrimination Against Women (CEDAW) which many countries, including Barbados, have ratified, organizations have created superficial policies which they label as "gender equal" (Dunn 2004). Many companies have created the post of "part time" worker to provide more employment especially for young women, and as Chioda (2011) notes, some have created a shift system which is said to be more flexible and accommodating to mothers. Through mechanisms such as contracts,[9] unequal relations of power become apparent as organizations determine the parameters of the job, shaping workers into the ideal worker subject who must abide by the work stipulations so their contract can be renewed. Dawn, one of the mothers in the study, speaks of a young woman who entered the university from the local community college with an "almost perfect GPA."[10] Her future appeared "bright" as she was employed part-time and studying full-time. This young woman became pregnant in her first semester at the university. Unfortunately, her workplace discovered she was pregnant, and her annual contract was not renewed. She had no legal recourse since she was not appointed[11] and worked on a part-time contractual basis. The loss of her job left her financially dependent on her boyfriend and unable to continue her university education which she believed would have allowed her access to more stable employment (Bailey 2009). It appears as if "motherhood entrenches women's low status and justifies their dependence on men" (Sistren 2005, 34), as men are better positioned to achieve "success" based on the patriarchal guidelines (Barriteau 2001). Amidst the international advocacy for gender equality, the personal experiences of women like Dawn are often ignored and illustrate the inefficiency of the policies and international conventions, and the perpetual unequal gendered relations of power; or more importantly, the experience demonstrates the ways in which gendered norms are reproduced within the context of privatized profit-driven political economies.

In some of the participants' experiences, motherhood became both an internal (personal reason) and external (organizational culture/structure) obstacle for women's career path. In Leah's experience below, motherhood appears to operate in an oppositional rather than supportive manner where she must choose family over job promotion, as if motherhood and paid work are linear and fixed. However, in analyzing the narrative, I argue that family and paid work are not linear but in fact often coterminously support each other, as seen in the interview. In negotiating their career and family life, some mothers make decisions based on the belief that careers impede good mothering. Leah, who postponed a promotion, admitted this saying,

> How my company is structured right now, promotion is probably a very far away off because I'm the only HR manager; there are no other HR managers. The only other promotion for me would be to become the

> group human resource manager, which is not something I wish to do right now. I'm happy where I am, where I do what I have to do, I manage my deliverables, I manage how I deliver my deliverables, and I move on. As opposed to having the responsibility of, you know, not being able to say "no" to anything, or say that I can't, because the ultimate "buck" stops with you as a group HR manager. My boss is of the opinion that once I get done what needed to get done, it doesn't matter how you do it, so taking a vacation day to him (shrugs); I said literally I am not coming in on Friday, I'm going to my son's sports. He was like ok, no problem.

Leah is a middle-class, tertiary-educated Barbadian mother of one. The structure of her workplace and her job position allowed her to engage in her son's school sports (and other activities, as she notes later in our interview) as long as she completed her work. This suggests that engaging in maternal practices is only permitted when work tasks are completed. Thus, there is a shaping of this female worker subject to gain maximum output, through a simultaneous regulation of the maternal subject. This is the complexity of motherhood and work to which I refer, where the relations of power within the workplace are gendered and used to shape the maternal and worker subjectivities. In Leah's experience her desire to be a "good mother" by being available to attend her son's school activities was a motivator to perform well at work.

This desire, however, also discouraged her from advancing her career, as she believed it would not be possible to be a good mother based on the demands of a higher job position. Despite perceived actions of transgression (e.g. Leah's decision to attend her son's events after completing her paid work), normative discourses around work and mothering places the worker subject at the forefront in the place of work, allowing for the presence of the maternal subject in a somewhat regulating capacity for the worker. Thus women, specifically mothers, are coerced into becoming gendered subjects. The promotion of the concept of the ideal worker as investing time within the company (a practice encouraged by capitalism's desire for maximum profit) prevents the advancement of many mothers within the institution/organization. Even when promotions do occur it does not provide the kind of financial stability promised. In fact, one mother (Joan) noted that despite the great amount of time and effort she spends at work, she still could not make enough money to support her household and must depend on her husband for "financial stability." Thus, even when great time is invested in the job, it does not always affect career or financial advancement, especially as the objective is aligned with neo-liberal beliefs of making profit, often translating into minimum and low wages for employees (Giroux 2008).

Using the Maternal to Shape the Worker Subject

The mothers, who engage in income-earning activities which require them to interact with children/students or in one case, manage employees (jobs that are care work oriented[12]) and argue that their skills are honed by their experience of motherhood. This argument is not necessarily transgressive, neither are the behaviors stemming from within this normative maternal discourse of the caring, emotional and affective mother. Joan, a married mother and teacher, described this stating,

> Ummm, well, the nursery part of it wasn't difficult to me, but I find like now that he is in like a primary school setting now, it's a little harder because I find there's a difference, like I get a lot of papers, things like what I would send to my parents, I am now getting those letters from the school and consent forms, and sometimes I am like where is the consent form; so I guess for me it's now that I am experiencing what my parents may have experienced. So it is a real eye opener for me, and it has shown me that sometimes, before I had Josh, and if I sent home something to be signed and it never came back or it came back late, I use to say "this parent is something else you know," but now I am understanding that there is so much that a mother has to do, to remember, to put in place, that there comes a time where some things are misplaced and you honestly forget to send in those forms.

This comment contains within it some of the underpinnings of the judgment which women administer to fellow women, a practice Ruddick (2007) calls "maternal abdication," and Foucault (1977) refers to as surveillance. While Foucault's writing does not speak directly to women, his narrative around surveillance among other groups such as people who are incarcerated can be juxtaposed to illustrate how "mother blame permeates much of our culture" (Baker 2004). The perspective that mothers are to be blamed homogenizes all mothers, denying them the multiple experiences of motherhood. Hence, it was only when Joan's son started primary school that she began to empathize with the mothers of the students in her class, and as such, her interactions with them became more sociable and the lines of communication became more open. Joan's experience shows how perception can influence interaction/relations and how experiences can change relations of power. Before becoming a mother, Joan had a romanticized perception of motherhood which affected her expectations of mothers generally, but specifically those whose children she taught. Changes in her perception of mothering came from her experience as a mother and recognition that the judgment on other mothers was not fair. However, Joan's mothering became framed within the discourses that outlined its incongruence with work. Lorde (1984) would argue that Joan became "empowered to action" as she began to live from inside out; in other words she became in touch with the power within herself, informing the world around her. Still, this changed perception and new understanding did not appear to be deep enough to cause a holistic change in Joan, but her experience made her a more effective teacher, in that she formed stronger home/school relationships, thereby providing the students with a "holistic experience" and gaining the "best" from them.

Still, Joan's realization is framed within normative discourses on mothering which naturalizes mothers as caring and emotional, especially as Joan's work as a teacher is primarily care oriented. The framing of empowerment to action in this way, positions maternity as foremost, and reiterates the naturalization and all-encompassing characteristics of motherhood, which now creates a more effective worker subject. In fact, Joan notes that it was especially beneficial to her career as she is employed in a private school and thus parents' satisfaction is pertinent to the school's growth. More importantly, these caring attributes gave Joan a feeling of satisfaction in "herself," although this is in relation to her maternal and worker practices. The use of these skills illustrates how the maternal and worker subjectivities become conflated for the benefit of the organization. While it is true that there are other non-mothers on staff who may be performing very well, according to Joan, her class outperformed all other classes in fundraising activities, and the parents of her students have been very supportive in assisting her in any school affair. This is not to negate the other dynamics that may impact the relationship between Joan and her parents, such as her personality as well as the age of her students,[13] but it shows how her maternal subjectivity changed her teaching practices and relationships in a manner that benefited her place of employment.

Like Joan, Sandra's experience illustrates how her maternal skills and knowledge are harnessed by the workplace to enhance work output. She explains how her attitude to work improved, noting that it was not negative when she was a non-mother, but she became more understanding and empathetic as a mother,

> … I find that I am, not that I wasn't a caring teacher but I find that I look out more for children now that I understand what it's like to be a mother and the little issues that the children have, you know I take that a bit more seriously now; and, I find that you become more motherly then to children not saying that I used to be some big rough person that didn't used to take on children…. yeah it's different and then when parents air their concerns you could actually relate now to what they are saying like they say well my daughter am, didn't want to come school this morning, you know some little frivolous excuse I now can relate to this cause my daughter does the same thing on mornings telling me mummy my feet are broke and I can't go to school on mornings, my feet are broken… I can't … you know parents would come in with these excuses and I'll be like, for real, they for real they got to be telling lies on these children cause can't be telling me … and it's only now that I can appreciate when their parents come to you with these issues with their children quite honestly.

This 32-year-old single mother and teacher is admitting that in becoming a mother she "look[s] out more" for her students and has become more "motherly." When Sandra speaks of how her motherhood affects her interaction with her students, I think of this channeling of her love to her students as what

Cheng (2004) referenced as "diverted love,"[14] that is, love given to another child to pacify the absence of one's child. However, in this context it is not done to appease the absence of one's child as noted in Cheng's (2004) study of transnational mothers, but rather, it is seen as an act of transgression that improves the teacher's interaction with her students. Although Sandra has associated care and understanding to maternity and has thus subscribed to the discourse of mothering as caring and affective, Lorde (1984) would argue that it is more than the act of just being motherly, but rather it is what being a mother forced her to do, that is, to feel deeply and thereby tap into the power of her erotic, that is important. However, this will only be so if she has embraced all aspects of her identity including the self (Lorde 1984). But as seen in the discussion so far, the maternal and worker subjectivities have consumed all other identities. Therefore, I would agree with Lorde that the focus is really on what the maternal subjectivity has forced her to do (to be excellent). But I would diverge and note that it is not transgressive but rather transforms the worker subject into one that is ideal for the neo-liberal work environment, one that is regulated.

Sandra's assertion that now she can "appreciate" and "relate" to the mothers of the students in her class speaks to an affective component that is engrained and naturalized in perceptions of mothering. Firstly, upon becoming a mother her interactions with others were altered almost immediately; and secondly, maternal discourses are used as markers for surveillance of other mothers with whom Sandra interacts. Similarly, Berlant (2011) notes that the optimistic attachments one has to certain ideals have emotional connections that sustain them even when proven to be unachievable. This maternal subjugation consists of self-regulatory behaviors but also regulates the behaviors of other mothers. Sandra's thoughts describe how mothering is shaped not only by the subject being interacted with (children), but by maternal skills and ideals that have been internalized and attached to emotions.

Simultaneously, she admits that having an ill child can distract a mother from her work, but this pressure can be used positively as it forces one to "be on task". According to Sandra,

> … you have to make sure that ***you're always on task for sure***, because ummm, if she … if something happens in the morning time at home I find that, you take it on depending on what it is you can take it on, like if she's ill or something, you take it on you come to work and you just can't focus, you're more focusing on if the child … how its feeling, what she's doing right now, um those sort of things so yes it … it does affect, it does affect. [emphasis mine]

Therefore, the argument noted by Crosby (1991) and Glenn et al. (1994) that having a child can distract an employee from the objectives of the organization is technically accurate, but this does not result in "inefficiency" as implied. In fact, being aware of the possibility of disturbances persuaded mothers to do what was necessary to maintain work output, and this often resulted in a greater output. Thus, this maternal subjugation can be a motivational force for the worker. Furthermore, like Joan, Sandra believes that her experience as a mother enhanced her rapport with her parents on the job,

> … but positively though you know you become more nurturing as a teacher towards students that's the honest truth and not only the, the academic side you look after them in other ways like, you can't bear to see a child sick or umm hungry that's the honest truth and you try to contact that parent as fast as possible. Before I was like, "well that's their problem," you know, …but now you would call the parent and you would actually want to find out exactly what transpired, why is it that this child didn't have this or didn't have that if its financial, the child was being punished, what's the reason the child doesn't have x, y or z. You know you're more, I find that you're a better person all round, that's the honest truth. You're tired, more tired than usual, but you're a better person all round.

Here again we see the discourse of the mother as nurturer emerging, but we also notice that this achievement appears to give Sandra a personal sense of satisfaction and power. Interestingly, when Sandra says she is a better person all around, this "person" to which she refers is the worker subject as seen in the reference to what has improved (more awareness of the children's personal needs, better communication with parents). This sense of satisfaction illustrates the institutionalization of motherhood[15] as care skills become associated with mothering. These practices are presented as naturalized for the worker subject because she is a mother. What is clear is that for the women in this study, their maternal subjugation shapes their perception and changes their interaction with the work environment. Kim, a young non-Barbadian national lawyer and mother of one, agrees with Sandra and brings out the benefit of being a mother when she compared mothers to non-mothers saying,

> I think they (non-mothers) can just spend more time at work. Like my secretary, she might come in on the weekend and stuff like that. I guess work becomes a lot more of your focus. I guess one thing that probably does change is your perspective… you probably handle stress a lot better because work isn't like the sole or most important thing in your life. Something happens, you just, and you can see it for what it is. You have perspective about it; it's not like the biggest crisis in your life. I guess that's probably one thing: I've become more laid back when it comes to a crisis. My secretary would be freaking out and I'd be "relax" you know, "We'll fix it somehow. Don't worry." So I guess I've become a lot more laid back about that. And I guess probably relationships; once you have a child maybe your interpersonal relationship skills get a little better because we get a little more empathetic…

The above assertion outlines a mother's expectation of someone who does not have children. It hints to the characteristic of the ideal worker as one without familial and social reproduction responsibilities. This norm reinforces the care of children to the mother, abdicating fathers' responsibilities in the process. Kim draws upon maternal discourses which suggest that mothering changes one's perspective and priorities of life, but in doing so she illustrates how her maternal skills are used to manage work stresses, and how both the maternal and work subjectivities operate coterminously to shape each other. Kim admits that the challenges she encounters as a mother have made her more relaxed and better able to deal with the stresses of her job. Her reference to perspective speaks to her positioning of other circumstances in relation to motherhood, illustrating the coexistence of the maternal and worker subjectivities, and blurring the private/public space.

Furthermore, Kim reproduces the assumption that motherhood is the principal marker of womanhood and should be the most important aspect of one's life. She assumes that if a woman does not have children, she does not have anything else that is meaningful in her life, she only has work. There is a supposition by Kim that this disadvantages non-mothers as they are stressed out by work-related things, whereas a mother would not be, because she knows that things could be worse: that is, rather than something going wrong at work, something could be wrong with her child. While on the one hand she may be countering the narrative of unproductive mothers at work for whom mothering distracts them from their jobs, her counterargument relies on a devaluing of the lives of women without children, seeing them as without a life outside of work and without perspective on what matters most in life. In addition, she fails to account for the hierarchical relations of power between her and the secretary. This may better explain why the secretary is "freaking out" when there is a work crisis, rather than the fact that she has children. After all, Kim, the attorney-at-law, is the one with the power to determine what is a crisis or not. Kim's suggestion that while women who do not have children can spend more time at work though they approach work without the kind of emotional maturity that mothers have demonstrates how all women are disciplined by motherhood as an institution and are defined by and against it.

Alternatively, the skills or effect of work can also impact the family. Rosanne, a single non-Barbadian mother explains some of the challenges she faces:

> … With him, he is not a dumb child but he is lazy. So a lot of times I realize I cannot trust him to take down his information properly in terms of (if) you have homework write it down on this book so when you get home I can check on it. If I don't go looking for him on an afternoon he would be in the yard running around and playing and no amount of scolding, quarreling, beating, had really changed that. Or if he comes to the class to sit down to get his homework done, he would tell you he is going to the bathroom and then (laughing) you just don't see him. So I still have to end up having him to do the homework on an evening when

> we get home probably after 6. It's just drama at that point because he's tired; he's just had a whole day of school. You don't necessarily want to necessarily sit down and do schoolwork by the light of an electric bulb. Ummm, there are times when specifically for him and with this job here in Barbados because I'm not mobile, it is difficult in terms of getting to school at a decent hour, probably giving him an extra hour to sleep or something. Ummm, last year there were a lot more things I had to do at the school in terms of getting organized and stuff so on a Saturday I would meet at the school as well.

Rosanne's description of dealing with her son in the school environment gives an idea of the intimate struggles of mothers. We see the heterogeneity in her experience because of the confluence of her location (not in her home country), type of mother (single), address (not on the bus route), economic position (minimum wage), and social circumstance (e.g. she is studying; has no familial support). The absence of an extended family unit means that Rosanne must rely on friends; but according to her, this is very difficult as the society is one where "friends are merely friends when they can get something from you." To combat this, Rosanne has implanted herself in the "church family." Thus, her Christian association not only offers tangible benefits but also impacts her maternal practices.

Furthermore, Rosanne's limited time is infringed upon by her restricted mobility. Her financial circumstances prevent her from purchasing a vehicle, and the location of her home requires that she walks at least a quarter mile to get one bus, and then another quarter mile to get a bus that will take her to "work". In addition to this being physically taxing, it depletes the limited time Rosanne has, and it emotionally frustrates her as she sees the loss of time as preventing her from being a "good mother" as well as negatively affecting her son. Rosanne notes that lately her son tells her, "but I am speaking to you. You are listening but you're not listening." She noted that this comment made her rethink her actions and prioritize personal quality time with her son. Here we see how ideas of what mothers should do (spend quality time with their child but also provide financially) conflict with each other. Even as this occurs, Rosanne navigates by reshaping her mothering because of the affective attachment to the outcome of a "loved and well cared for child."

Rosanne's story corroborates the experience of many Caribbean women over the decades as it shows that while childcare can be transferred to someone else temporarily (teacher, nanny, grandparent), mothers believe they should never stop thinking of their child (Anderson 1986; Mohammed and Perkins 1999), which illustrates how motherhood is a priority in their lives.

On the other hand, there is a view which highlights the benefit of work skills on the family. Leah, in explaining how her job affects her role as a mother, states,

> … Umm, I would say yes (it affects my role), because with my job I do a lot of talking to people, so I do a lot of counseling, directly/indirectly, professionally/personally, um I do discipline employees, obviously by the

> code of discipline that the company has in place, but because of that I do a lot of talking with J,[16] or I try to do a lot of talking, so I try to explain why I say something or why I do something the way I do, um, so that he would understand that, so I think because of my personality, one, and because of my job, that's the approach that I take with him.

Leah is a Human Resources Manager, and her son is older (9 years old) than the children of Joan, Kim and Sandra, who also share the belief that their jobs affect their mothering. This difference in age means that Leah's son can better understand the verbal discipline she uses. Leah demonstrates how her job skills are woven with her personal attributes when she says that it is the interaction of her personality and job training which create this approach to disciplining her child. Additionally, unlike the findings of Evans and Davies (1997) which purport that Caribbean parents do not use reasoning and explanation to discipline children thereby resulting in "negative socialization" at times, Leah practices and believes that a parent should explain the particular discipline to the child so that they truly understand. This is a changing maternal perspective impacted by agencies such as UNICEF who advocate for children's rights (ECLAC 2011).

Historically, Caribbean parents' disciplinary method has been corporal and authoritative, often silencing the child's voice (Barrow and Ince 2008; Anderson 1986; Evans and Davies 1997). However, with the Convention on the Rights of the Child (1989) and global advocacy for children's rights, there have slowly been changes in the culture of parenting in the Caribbean (Barrow and Ince 2008). It is further corroborated by the education system as Leah's university education and her job training as a human resource manager enlightened her about the benefits of reasoning with the child which includes justifying the punishment to him, allowing him to explain his reasons for behaving in a particular way, and indicating her ultimate goal to the child. Nonetheless, Leah grappled with the oppositional positions of her cultural expectation of parenting and her educated expectation. While she admits that her way of disciplining may be "a little bit too soft," she believes that her son is not at a disadvantage because of it, especially since her husband "would clamp down on him anyway." Thus, Leah relies on her partner as a "backup" in disciplining their son.

Hence, Leah unwittingly does two things. Firstly, she advances the flawed gendered discourse that mothers are the main parent with fathers being available for the economic and secondary disciplining. While historically this gendered division of labor was the norm in the Caribbean when fathers were present in the household, this is slowly changing. Thus, Brown et al.'s (1993) study in Jamaica indicates that fathers are quite involved in the parenting of their children doing more than just disciplining and financially providing (although still unequally). While Leah's experience reflects Brown et al.'s findings, it is not as transgressive. Many of the mothers, like Leah, mention that their husband "helps" in other care practices. Their perception of fathering is framed within gender norms which places fathering in a secondary position in the parental relationship: as help. Secondly, while Leah's experience depicts transgressive practices ("I try to explain why I say something or why I do something the way I do," rather than the historical way of blind obedience by the child or corporal punishment), it still reflects the instructions and objectives of the institution of motherhood, in that her choice is based on the advice and recommendations of government and NGOs. These institutions regulate the practices of the maternal subject, thereby reinforcing the gender norms that homogenize and naturalize motherhood.

Conclusion

It is clear from the narratives above that the women interviewed for this study perceive work as money-earning activities, but also label their social reproduction as work. In fact, work is presented as instrumental and simultaneously subjugating, regulating, a source of discrimination and at times as transgressive. While they perceive mothering and work as different, almost oppositional actions, their experiences illustrate the interconnectedness of these endeavors. This chapter demonstrates how the maternal subject produces and regulates the worker subject, which simultaneously places it in a position of "Other." While the "means to an end" relationship between the maternal and worker subjectivities is obvious, the benefit brought to each sphere as a result of the other, is undeniably substantive, and points to the non-linear lines of demarcation. Despite the varying careers of the women, each purports that being a mother has made them more caring, enhanced their communication skills, improved their planning skills and efficiency, and made them better at multitasking and dealing with stressful work situations. They failed to identify these skills as already existing in their mothering, as already being honed through their negotiation and navigation of motherhood, thereby merely being used by their workplace. Their experiences demonstrate a neo-liberal ideal in that a set of qualities that are "care" attributes, which are connected to the subject of "mother," are utilized to enhance and profit the workplace. The participants' narratives are framed within the normative discourses of motherhood as affective, emotional, and caring. They, therefore, reproduce the maternal subjectivity that is used to produce the ideal worker subject in a neo-liberal context.

Notes

1 Driven by the goal for regional integration and the aftermath of an earlier attempt through the Caribbean Federation in the 1950s, Caribbean islands ratified the Treaty of Chaguaramas pledging support for integration. One of the means through which this was enacted was the Caribbean Single Market and Economy (CSME) which allowed the free movement of skilled CARICOM nationals to work and reside in other CARICOM territories. As a member of CARICOM, Barbados signed on to CSME, allowing skilled citizens of CARICOM states to work and reside within its shores. Barbados, having one of the better-paid public services in the Eastern Caribbean, became a hub for many skilled CARICOM nationals. Using Barbados as a case study, I captured the experiences of non-Barbadian# and Barbadian women residing in the country.

2 The experiences shared in this chapter are taken from a larger study on women's negotiation and navigation of mothering and work in which 21 Caribbean women residing in Barbados were interviewed. All names have been changed to ensure the confidentiality of the participants in the study.
3 In this context, "take the dog treatment" means the inhumane, abusive, poor treatment and conditions of her work environment.
4 This belief is seen in the work by Engels, *The Origin of the Family* (1973). It is discussed in Inge Bretherton's 1992 essay, "The Origins of Attachment Theory: John Bowlby and Mary Ainsworth," *Developmental Psychology* 28.
5 I use the term "institution" from a Foucauldian premise where it is a system of regulations, but it is also a space where motherhood is produced and reproduced, enacted and acted upon, and where maternal subjects are formed. Mothers are acting on this institution and not merely produced by it.
6 I refer to this concept "good mothering" to speak to the institutionalization of mothering as a set of ideas of what "good" mothering entails.
7 Work in this section refers to their employment activities.
8 Sistren is a collective of Caribbean women activists.
9 Contracts stipulate the length and hours of the job and if any benefits will be included. For example, the employer does not pay any benefits or taxes for an independent contractor. Employees also do not have to explain or justify contract non-renewal.
10 Grade point average
11 In Barbados, the term appointed refers to permanent appointment in which the employee is expected to continue in that employment until retirement.
12 As conceptualized in Trotz and Peake. 2001. "Work, Family and Organising: An Overview of the Contemporary Economic, Social and Political Roles of Women in Guyana," *Social and Economic Studies*, 50(2): 67–101.
13 Students are in the 6–7-year-old age group; this is also the class where the students write the first national assessment exam in Barbados called the Criterion Referenced test. These tests are conducted twice during the child's primary school years (6–7 years – Infants B, & 9–10 years of age – Class 2).
14 Cheng uses this term to describe the channeling of love through the care labor of women of color to their employees' children and families.
15 Institutionalization of motherhood refers to how it is implemented/operationalized.
16 Child's name was removed for confidentiality reasons.

References

Acker, Joan. 1990. "Hierarchies, Jobs, Bodies: A Theory of Gendered Organizations." *Gender and Society*, 4(2): 139–58.

Ahmed, Sara. 2010. "Happy Objects." In *The Affect Theory Reader*, ed. Melissa Gregg and Gregory Seigworth, 29–51. London: Duke University Press.

Anderson, Patricia. 1986. "Women in the Caribbean (Part 1)." *Social and Economic Studies*, 1: 291–325.

Antrobus, Peggy. 2004. *The Global Women's Movement: Origins, Issues and Strategies.* London: Zed Books.

Apple, R.D. 1995. "Constructing Mothers: Scientific Motherhood in the Nineteenth and Twentieth Centuries." *Social History of Medicine*, 8(2): 161–78.

Apter, Terri. 1993. *Working Women Don't Have Wives*. New York: St. Martin's Press.

Baker, Christina. 2004. "Telling Our Stories: Feminist Mothers and Daughters." In *Mother Outlaws: Theories and Practices of Empowered Mothering*, ed. Andrea O'Reilly, 95–104. Toronto, Canada: Women's Press.

Bailey, Barbara. 2009. "Gender and Political Economy in Caribbean Education Systems: An Agenda for Inclusion," *Commonwealth Education Partnerships*. Accessed 11 February 2021. https://www.cedol.org/wp-content/uploads/2012/02/102-105-2009.pdf.
Barbados Integrated Government. 2023. "Demographics". Accessed April 13, 2023. https://www.gov.bb/VisitBarbados/demographics#:~:text=Ethnic%20groups,white%20and%201.3%25%20South%20Asian.
Barriteau, Eudine. 2001. *The Political Economy of Gender in the Twentieth-Century Caribbean*. Hampshire, England: Palgrave MacMillan.
Barrow, Christine, and Martin Ince. 2008. "*Early Childhood in the Caribbean*." The Hague, Netherlands.
Berlant, Lauren. 2011. *Cruel Optimism*. Durham: Duke University Press.
Bretherton, Inge. 1992. "The Origins of Attachment Theory: John Bowlby and Mary Ainsworth." *Developmental Psychology*, 28: 759–75.
Brown, Janet, Patricia Anderson, and Barry Chevannes. 1993. "A Contribution of Caribbean Men to the Family: A Jamaican Pilot Study." The Caribbean Child Development Centre School of Continuing Studies, UWI.
Bush, Barbara. 2010. "African Caribbean Slave Mothers and Children: Traumas of Dislocation and Enslavement across the Atlantic World." *Caribbean Quarterly Slavery, Memory and Meanings: The Caribbean and the Bicentennial of the Passing of the British Abolition of the TransAtlantic Trade in Africans*, 56(1/2): 69–94.
Cheng, Shu-Ju Ada. 2004. "Right to Mothering: Motherhood as a Transborder Concern in the Age of Globalization." *Journal of Association for Research on Mothering*, 6(1): 135–44.
Chioda, Laura. 2011. *Work & Family Latin American and Caribbean Women: In Search of a New Balance*. Washington, DC: The International Bank for Reconstruction and Development.
Crosby, Faye. 1991. *Juggling: The Unexpected Advantages of Balancing Career and Home for Women and Their Families*. New York: MacMillan Inc.
Dunn, Leith. 2004. "Women and Work: Policy Implications of ILO Conventions." In *Gender in the 21st Century: Caribbean Perspectives, Visions and Possibilities*, eds Barbara Bailey and Elsa Leo-Rhyne, 303–32. Kingston, Jamaica: Ian Randle Publishers.
ECLAC. 2011. "Eclac and Unicef Examine Parental Leave in the Region," news release. Accessed 11 February 2021. http://www.cepal.org/en/pressreleases/eclac-and-unicef-examine-parental-leave-region.
Engels, Frederick. 1973. "The Origin of the Family." In *The Feminist Papers from Adams to De Beauvoir*, eds Alice S. Rossi, 480–95. New York: Columbia University Press.
Evans, Hyacinth, and Rose Davies. 1997. "Overview Issues in Childhood Socialization in the Caribbean." In *Caribbean Families: Diversity among Ethnic Groups*, eds. J.L. Roopnarine and J. Brown, 1–24. London: Ablex Publishing Corporation.
Firestone, Shulamith.1997. "The Dialectic of Sex." In *The Second Wave: A Reader in Feminist Theory*, ed. Linda Nicholson, 19–26. New York: Routledge.
Foucault, Michel. 1977. *Discipline and Punish: The Birth of the Prison*. Translated by Alan Sheridan. New York: Vintage Books.
Giroux, H.A. 2008. *Against the Terror of Neoliberalism: Politics Beyond the Age of Greed*. Boulder, CO: Paradigm Publishers.
Glenn, Evelyn Nakano, Grace Chang, and Linda Rennie Forcey. 1994. *Mothering: Ideology, Experience and Agency*. New York: Routledge.
Grint, Keith. 1991. *The Sociology of Work: An Introduction*. Cambridge: Polity Press.

Holmstrom, Nancy. 2003. *Introduction to the Socialist Feminist Project: A Contemporary Reader in Theory and Politics*. New York: Monthly Review Press.
Lorde, Audre. 1984. *Sister Outsider: Essays and Speeches*. New York: Crossing Press.
Mohammed, Patricia, and Althea Perkins. 1999. *Caribbean Women at the Crossroads: The Paradox of Motherhood among Women of Barbados, St. Lucia and Dominica*. Kingston: Canoe Press University of the West Indies.
Mondesire, Alicia, and Leith Dunn. 1995. "Towards Equity in Development: A Report on the Status of Women in Sixteen Commonwealth Caribbean Countries." Georgetown, Guyana: Caribbean Community (CARICOM) Secretariat.
Nash, Terre. 1995. "Who's Counting? Marilyn Waring on Sex, Lies and Global Economics." National Film Board. Accessed 13 July 2015. https://www.nfb.ca/film/whos_counting/.
Reddock, Rhoda. 1994. *Women, Labour and Politics in Trinidad and Tobago: A History*. London: Zed books.
Ruddick, Sara. 2007. "Maternal Thinking." In *Maternal Theory: Essential Readings*, ed. Andrea O'Reilly, 96–113. Bradford, Canada: Demeter Press.
Sistren. 2005. *Lionheart Gal: Life Stories of Jamaican Women*. Jamaica: University of the West Indies Press.
The World Bank Group. 2023. "The World Bank Data – Population, Barbados". Accessed April 13, 2023. https://data.worldbank.org/indicator/SP.POP.TOTL?locations=BB
Trotz, D. Alissa, and Linda Peake. 2001. "Work, Family and Organising: An Overview of the Contemporary Economic, Social and Political Roles of Women in Guyana." *Social and Economic Studies*, 50(2): 67–101.
Weber, Max. 1992. *The Protestant Ethic and the Spirit of Capitalism*. New York: Routeledge.
Weeks, Kathi. 2011. *The Problem with Work: Feminism, Marxism, Anti Work Politics and Post Work Imaginaries*. Durham and London: Duke University Press.

5 Psychosocial Uncertainty

Making Sense of Institutional Suffering in Trinidad and Tobago

Leslie Robertson Foncette

Introduction

In this chapter I use an intersectional analysis and integrate Black and Caribbean feminist frameworks to examine how people make sense of structural failures when gender, race and class are limiting factors, particularly for poor and working-class Black women in Trinidad and Tobago (T&T). Focusing primarily on how women navigate maternal and reproductive care, and drawing connections to other routine survival struggles, I explore what happens when people meet challenges or misfortune engaging with everyday public services. Using incidents described in news media, personal experiences and accounts of people living in T&T, I contextualize this analysis in the history of colonialism, plantation slavery and the creation of creole society in T&T.

To inform my study, I draw from the work of sociologist Javier Auyero, and anthropologist Débora Swistun, who recount how people living in a shanty town in Buenos Aires, Argentina deal with the impacts of environmental disaster due to industrial pollution. Auyero and Swistun examine how people make sense of the disaster and cope in the face of toxic pollution, dispossession, physical harm, lack of enforcement of governmental regulations or accountability. They describe the confusion, distrust, and response to misinformation as toxic uncertainty, highlighting how systemic failures for a community yield a pollution of thought and confidence for the people who suffer. Examining parallels to the impacts of system failure in healthcare in T&T, I coined the term *psychosocial uncertainty* to describe how people cope when governmental systems of care fail them. Although psychosocial uncertainty is a phenomenon that can be applied to many areas of public service as well as private institutions that significantly impact routine life in the Caribbean, such as the banking sector, I limit my examination here to healthcare, particularly the experiences of women accessing maternal, sexual and reproductive healthcare.

In this chapter I define psychosocial uncertainty and explain some contexts in which it occurs in T&T. Drawing from a commonly used source of public information – newspaper articles – I explain how inefficiency or dysfunction and narratives of tragedy become a basis for public rhetoric on the state of

DOI: 10.4324/9781003143550-8

healthcare, shape day-to-day interactions and influence people's perception and decisions around care. I also discuss some of the research literature on perceptions of healthcare and how options are limited for women who lack resources. I draw a connection to the historical contexts of colonialism and slavery that undergird ideas of gender, race and class hierarchies in T&T. I explain why these contexts matter in the contemporary space, and their influence on how people are treated, and what options they have at their disposal. Next I discuss how my positionalities frame my understanding and perception of the situation and the limitations. I also consider how Black feminist scholars have contextualized the experiences of people too often denied access to institutional power and explain how it applies in T&T. I conclude by discussing some of the broader implications for how psychosocial uncertainty surfaces frequently, make inferences about how people navigate it, and consider a way forward.

Psychosocial Uncertainty

A common artifact of interfacing with public systems in T&T – those paid for with government funds or managed by governmental institutions – is that people become familiar with uncertainty: waiting for information, not fully understanding complicated and bureaucratic processes; waiting to access care or receive service. Uncertainty seems to manifest from lack of information or misunderstanding communication by those in positions of authority, be they the doctor, the nurse, or the security guard.[1] What should be basic exchanges, such as obtaining medical tests or a form of identification become complicated, bureaucratic processes coupled with uncertainty about how long one will wait to access service. Perhaps a general lack of faith in the effectiveness of institutions leads to a persistent state of not knowing that is uncertainty. The response is both internal and interpersonal, that is, there is a psychosocial component: people cope by adapting their work and social habits, for example taking a day off work for an early morning appointment because you expect it will take all day, or coming to work, but leaving while on duty to get something done as quickly as possible. Whether this practice is with the approval of supervisors or not, it is colloquially known as "running away" – a term that is used often in Trinbagonian[2] discourse. People adjust their expectations, and over time, their attitudes, and beliefs about efficiency. There is the feeling of surprise when transactions happen quickly, and they highlight anomalous success stories of efficiency or good customer service. However, this form of coping is fraught with confusion and frustration, perhaps because we have an innate desire to have clarity about routine activities. To mitigate the impact of waiting and uncertainty navigating bureaucratic systems people prepare for the worst-case scenario, for instance, arriving in the early pre-dawn hours to line up for a clinic appointment, using private services when affordable, preparing to be inconvenienced or bracing for a recalcitrant civil servant or medical or customer service personnel, finding out who you know so you can jump the queue, or get someone to advocate with an authority figure on your behalf. This means adapting to situations that are not ideal, adapting to uncertainty.

Uncertainty is something that develops gradually, not as a reaction to one particular or tragic event. In their book, *Flammable: Environmental Suffering in an Argentine Shantytown*, Auyero and Swistun (2009) conduct an ethnography of environmental disaster that occurred over many years and was caused by multiple actors, both state and private corporations. The residents of Flammable experience toxic uncertainty, that is, symbolic violence (Bourdieu 1991) that takes the form of environmental suffering and an unconscious imposition of dominance by various power structures. Borrowing from sociologist Charles Tilly's (1996) term Auyero and Swistun argue that the interaction of "'invisible elbows' of external power forces and of everyday routine survival struggles" (2009, 6) creates a form of suffering where people are not certain of the main actors in their oppression, nor can they specifically pinpoint the mechanisms by which their suffering began. Misinformation and day-to-day stressors make it difficult to make sense of the whole picture.

Thus, psychosocial uncertainty is a way of experiencing institutional suffering, where ordinary citizens interact with institutions and experience barriers, frustration, and various forms of oppression. It is a way of explaining the pragmatic individual impacts, the shifts in collective mindset and the emotional and social responses to confronting bureaucracy, particularly when one embodies marginalized identities such as gender, race or class. Societal notions of distrust in public institutions, and sometimes misinformation or confusion, may lead people to employ strategies to avoid structural challenges, for example, long waits for appointments for medical tests, having to go to multiple clinics to fill prescriptions (*Trinidad Express* 2013a) or multiple offices to obtain a signature or stamp on an official document (*Trinidad and Tobago Newsday* 2017). People rely on processes outside the bureaucracy that do not serve them if they have the wherewithal and economic or social capital to get ahead.

In Auyero and Swistun's description of environmental disaster, residents' routine survival struggles – sustaining employment, providing for their families, caring for children, managing health problems caused by living in a toxic environment, and challenges associated with poverty and environmental disaster are marked by toxic uncertainty. Where toxic environmental pollution is a disrupter, routines signal a continuity of life, they help to provide stability, "have an ordering effect. They orient and simulate action" (Auyero and Swistun 2009, 142) in overcoming the uncertainty because people continue to live their lives. I argue that outside the context of environmental disaster[3] routine survival struggles include the business of navigating bureaucracy to access public healthcare or education, obtaining official forms of identification and seeking protection from law enforcement or the judicial system. Rather than the toxic uncertainty people experience in environmental disasters, this uncertainty is both psychic and social. Confronting confounding procedures, lengthy waits for services, salient narratives of death or negative outcomes that fuel the idea that one cannot rely on the certainty of favorable engagement with public services and institutions of power (*Trinidad Express* 2013b) is a similar type of suffering, I posit, to that which happens when people navigate environmental disaster. And the perceptions of service and decisions people make about how to navigate care or survive inefficiencies or dysfunction are influenced by marginalizations of gender, race or class.

Auyero and Swistun (2009) describe the "confusion, mistakes and/or blindness regarding toxicity around them" and the "silent habituation to contamination" (Auyero and Swistun 2009, 4). They describe how toxic uncertainty starts in the midst of environmental disaster and a devolution of trust and efficacy by those who should manage the problem. I posit that psychological uncertainty is a byproduct of a system designed to maintain order through gender, racial, economic and class hierarchies. Uncertainty about whether an ordinary citizen, lacking money or social capital can complete routines of daily life, access healthcare, obtain identification or other goods and services constitutionally regulated or managed by the state is a constant reminder of subjugation for marginalized people. It applies pressure on those most harmed to continually resist, or to accept that they deserve unequal treatment. An everyday citizen without money or access can believe they deserve better treatment from the state, but they consistently receive messages to the contrary, resulting in psychosocial uncertainty. Although a critique of how the public health sector functions is necessary, analysis of perceptions of institutional functioning may elucidate the discrepancies between the healthcare system's communication of efficacy – and the level of trust people have in them.

Attitudes and beliefs about healthcare reflect a psychosocial uncertainty because distrust does not always reflect various forms of data and because the system inherits a colonial orientation it presupposes no interrogation. People may lament the situation but at the same time ask "what yuh go do?" They seek redress through private resources or connections, they stand up for themselves individually, but is there a reckoning with the sources of oppression or a means to strategically end the inefficiencies or miscommunication that creates suffering? For example, people often express distrust in doctors at public hospitals; however, they acknowledge that the doctors who treat them in private hospitals are employed at public hospitals and were educated at the same institutions as physicians who work exclusively at public facilities. They choose private options, incurring personal expense, even when there are anecdotes that evidence less monitoring and potential harm by negligent doctors, out of fear of dealing with the conditions of public facilities. They compare healthcare services and access to countries with significantly larger economies than T&T, the most frequent comparisons being with the United States, the United Kingdom and Canada, but seemingly without any insight into the massive health inequalities in those nations, particularly the United States, nor the ways in which the commercialization of healthcare and the absence of a public health system renders healthcare inaccessible to many there. Psychosocial uncertainty is a culture of skepticism that is characterized by doubt, distrust and resignation in the institutions that are designed to help and protect us. It can result in the privileging of private sector and foreign health services at considerable expense, the avoidance of services, the automatic assumption of inferior positioning, sometimes leading service users to make decisions against their own interests out of a lack of confidence that care will materialize.

Antecedents of Mistrust

To ask people in the Americas who have been historically mistreated, particularly Black people, to trust systems of power is to deny centuries of exploitation and to deny that we live in the afterlife of slavery (Sharpe 2016). Black trust is a calculation. It is contingent on self-determination in spite of oppression, foreclosed on the impossibility of their modern existence and the certainty of ancestral commitment to survival. Black people's estrangement with the state is embedded within the context of colonialism, enslavement and indentured servitude, where Black bodies have not typically been treated with humanity (McKittrick 2014; Turner 2017). One can never fully trust state power when it has been employed to infringe one's basic rights, when the wealth of an isolated few, often living in far-away places is brokered by exacting every last ounce of strength from the flesh and blood of human beings kidnapped and forced to labor in lands far from home.

Contextualizing Plural Society

Historical, political, economic and social contexts in T&T affect contemporary society (Bolland 2005). Race and class structures contextualize how social hierarchies privilege some and create suffering for others. Domination, conformity, stratification and resistance are intrinsic to the socio-cultural fabric of colonial society (Brathwaite 1971). Race and class stratification were defining elements of the history, social and political culture of the Caribbean and remain embedded in systems of neocolonialism and capitalism there today (Allahar 2003). Meaning was given to origin (whether European white or Creole white), and race and status (whether aristocrat, merchant class, middle class or laborer) remained a silent, sometimes commanding dynamic in social negotiations (Baranov and Yelvington 2003). Relations in Caribbean society were contingent on an outside authority governing a confluence of different and differently positioned people within. Barbadian historian and poet E. Kamau Brathwaite (1971) described Jamaican society in the late eighteenth century and up to emancipation, as a creole society, one where control presumed an entanglement with the colonial hold on power. People of diverse origins were caught between a metropolitan European power, on the one hand, and on the other hand, the rules, and mores of a plantation arrangement. This was a multiracial society stratified according to Eurocentric superiority and hierarchy.

> 'Creole society' therefore is the result of a complex situation where a colonial polity reacts, as a whole, to external metropolitan pressures, and at the same time to individual adjustments made necessary by the juxtaposition of master and slave, elite and laborer, "in a culturally heterogenous relationship".
>
> (Brathwaite 1971, 4)

Parallels can be drawn between Brathwaite's descriptions of Jamaica with creole society in T&T, a former colony with similar history. The inherent preoccupation with class and color in creole society continues to this day. Hintzen (2002) argues that while notions of impurity excluded white creoles from colonial power and privilege, it assured them that power and privilege are a nationalist construction. This positioning continued to be true post-emancipation in T&T. Indentureship provided its own "legitimating regime of exclusion" (Hintzen 2002, 96) and racial stratification among all other groups gradually created a hierarchy. Although previously outside the acceptability of Creole whites, Portuguese indentured laborers would eventually become the merchant class, along with Syrians, Lebanese and Jews, and post-indenture Chinese immigrants. Indians and creole Africans were at the lowest rung of society; however, creole Africans, who had spent more time away from the plantations, and people of mixed heritage, often of mixed African and European, or Chinese descent, ultimately formed the middle class in the post-independence period. By the period leading up to independence in 1962, although being Black had been previously associated with inferiority, Black people began to claim Afrocentricity and pride in ancestral heritage (Allahar 2003). Thus, race and class stratification undergirded most aspects of societal order in T&T, the vestiges of which, no doubt, persist today.

In this context the social (and political) impact of distrust and disaffection with public institutions is borne out of a relationship with racial and class privilege in a colonized space. Although T&T is an independent nation, understanding of authority and class stratification has historical precedence in plantation society and colonialism, designed to place more trust in Eurocentric and external systems of control and governance, and to give preference to those who are more closely aligned with those positionalities. As Nigel Bolland (2005) describes it:

> Socially constructed categories and identities of "black" and "white" were developed in relation to each other within a racial hierarchy shaped by particular historical social forces. Hence, 'whiteness' is not a kind of trait or characteristic apart from; blackness' but is a claim of superiority over 'blackness'.
>
> (182)

People in today's societal context who are marginalized or lack sufficient socio-economic and political capital (typically involving access to private or foreign resources, political and economic power, foreign connections) might not necessarily trust any official systems to serve them as a priority. In local parlance, marginalized people are less likely to "get through" – a colloquial way of describing good fortune – when it seems unlikely. Confrontation with institutional power as a function of the racial stratification systems for which colonialism laid the groundwork is a setting in which psychosocial uncertainty bears fruit. For women of lesser economic means, this effect may be magnified.

The Intersecting Axes of Gender and Gender Relations

Gender plays a specific role in marginalization as do race and ethnicity. "Gender relations encode and often mask unequal power relations between women and men, and between women and the state" (Barriteau 2001, 26–27). In her work, *The Political Economy of Gender in the Twentieth Century*, gender and public policy scholar, Eudine Barriteau (2001) underscores how "power and domination are at the roots of gender inequality" (27). And gender inequality, rooted in domination, means inequality in the material and ideological outcomes for poor women, particularly those who lack other hierarchical status. The inequalities built into contemporary gender systems render them unjust, and, as Barriteau says, "continuously reinforced in their structures and practices" (31). For instance, the seventeenth century focus on women's reproductive capacity as an argument for the abolishment of the slave trade, or commodification of enslaved women as workers and producers, or cultural shifts in public medicine that lead to the masculinization of public reproductive care (Turner 2017) are historical contributing factors to gender inequality. Therefore, systematic oppression of women in contemporary T&T society is rooted in historical asymmetry of gender relations, and ideologies that designate women, and women's material conditions as subordinate. As such, it is no surprise that distrust, confusion and uncertainty develop for women of limited means who experience public healthcare systems.

Reproductive health is an essential aspect of women's health and a human right that is often subject to the structural limitations imposed by institutions and culture in society. Examining the perceptions, attitudes and beliefs women in T&T have about their options for care, and their efforts to navigate or adapt highlights how women respond to these structures. Patriarchal ideas, the over-representation of men and under-representation of women in government, leadership and decision-making roles across the world mean that most rules and decisions about reproductive healthcare are made by men. Moreover, crafting reproductive health legislation often happens without the input of women and often overlooks consideration for the hardships and challenges of the most marginalized and under-represented women. Consequently, gender inequality concerns important issues to consider in women's healthcare.

What Shapes the Experience of Uncertainty

Considering how patriarchal structures compound race and class inequalities, women may experience more uncertainty, especially women whose material conditions are more constrained. For instance, in an early study examining health equity through confidence and satisfaction in public healthcare Rudzik (2003) found that women in Trinidad[4] stated a preference for private over public healthcare. Additionally, women with higher incomes and who had recently used private care had less favorable perceptions of public clinics, perhaps because they experienced facilities, or patient treatment that was more timely or accommodating than that of public care. Private goods and services, those we seek in exchange for payment, are presumably circumstances that we can control. "The best that money can buy"; "money talks"; "money makes the world go round" are common sayings that belie our understanding of the power of money. As Rudzik (2003) stated "confidence of patients in public healthcare in Trinidad and Tobago is conditional… patients do not seem to *feel they can rely on public care* for complex cases" (248) [emphasis mine]. As such gender, class economic and social capital mediate success in navigating institutions and like toxic uncertainty, psychosocial uncertainty is the product of "puzzling and contradictory meanings (citizens) ascribe to it" (Auyero and Swistun 2009, 4).

Historical accounts of slavery in the Caribbean evidence how the commodification of medicine in capitalist healthcare systems, and hegemonic masculinity have influenced medicalization of childbirth, and patriarchal perceptions about how women should approach their reproductive care (Turner 2017). Consequently, the last two generations of people born in Trinidad and Tobago, like many industrialized and developing countries, have been born in hospitals rather than at home (United Nations 2012). My Trinidadian grandmother and her seven siblings were born at home in the 1920s and 1930s, as I believe was her Barbadian mother, my great-grandmother. However, my grandmother gave birth to all her children at a private maternity hospital from the beginning of the 1950s until her last one in the mid-1960s. Staffed by midwives and physicians, this medicalization of childbirth is considered progress. Although nurses and midwives still command authority on labor and delivery wards, they operate within the confines of a very masculine and hierarchical hospital structure, where orders are given from above by physicians, to be carried out by nurses, who are most often women. I consider Glenn's (1992) framing of race and gender hierarchies in the medical field and how those inequalities are applied in contemporary T&T. Although it is a plural, post-colonial society where people of African and Indian descent are the dominant groups in number, T&T is, as social anthropologist M.G. Smith describes plural societies, culturally split, "governed by dominant demographic minorities whose peculiar social structures and political conditions set them apart" (M.G. Smith 1984, 29). Smith argues that race and pluralism operate in T&T as a more complex form of hierarchical plurality because race, religion, culture, size relative to the whole population, political status, economic interests and other elements are factors that serve as a means of division or segmentation. Consequently, analyzing through an intersectional framework, hierarchies operate when women of different racial, ethnic and class positions engage the system to access care, seek information and seek redress. Although we do not have information on racial disparities in medical professions in T&T, we can consider signals of race and class stratification and inequality across time and domains. For instance, Brereton's (1979) study of race relations between 1870 and 1900 discussed discrimination against non-whites in business, the civil service, churches, courts, and schools; wealth, exclusionary organizations and large monetary prizes used by the middle and upper classes to control masquerade in T&T carnival (Liverpool 2001). Further, Natasha Barnes (1997) argues the Carnival Queen competition (from which non-white, and subsequently darker-skinned women were historically excluded) with extravagant prizes sponsored by the elite showed the stark disparities and "visceral disenfranchisement of people of color" (287). It is no surprise then that disparities privileging groups of higher socio-economic status and racial or ethnic privilege in the sphere of carnival reflect a broader sense of inequality and discontent that became magnified by and sparked social unrest in the nation and political movements lead by Black university students during the 1970s (Liverpool 2001). Using these frames gives a context to how poor women of color, particularly African or poor East Indian women in Trinidad and Tobago are likely to experience health and reproductive care in the public health system.

Sociologist Patricia Hill Collins (1986) offers a framework that shows how knowledge reflects the standpoint of its creators. Where structures of knowledge are oriented in a colonialist, white Westernized way, knowledge validation is Eurocentric, and tied to US or European power relations. Consequently, the interests of Eurocentric, colonial power pervade the themes, paradigms, and epistemologies of traditional scholarship, routinely distorting or excluding Black women's experiences from what counts as knowledge. This process is actualized via epistemological choices about whom to trust, what to believe, and why something is true or not. Collins argues that these choices at their core affect what truths will be believed and what knowledge is privileged. Therefore, epistemology frames how power relations shape our certainty about our lives, whose narratives are believed, and how that knowledge is formed.

Maternal Health as a Reflection of National Health and Development

Conversations about maternity outcomes and perceptions of maternity care were how I initially considered psychosocial uncertainty. Public discourse, provocative newspaper headlines (Webb 2017; Loutoo 2017; Wilson 2018; Ali 2019) describing maternal deaths, private conversations with friends and acquaintances about their choices for care, and publicly available information on maternal mortality (Roopnarinesingh, Ramoutar and Bassaw 1991; Hogan et al. 2010; Namasivayam et al. 2012; Blaize-LaCaille 2018; Hoyert 2022) suggested a problem. I was initially struck by some perceptions and attitudes about maternal health outcomes in T&T and the comparisons with the United States because maternal mortality rates in the United States are increasing rather than declining, and its rate is double that of other high-income nations (Kassebaum et al. 2016; Hoyert 2022). Moreover, there are vast health disparities for women of color, particularly Black women (Hosseini 2019; Jolly 2021; Villarosa 2018). In a majority Black and East Indian country like T&T the perception of foreign (North American, British, European) and local private medical settings (which lack regulation and structure of the national health system) as preferred was significant enough to warrant interrogation. Thus, maternal health outcomes, though not the only site for examining engagements with institutions, are important in discussing health in this chapter. They reveal how people make sense of health and well-being relative to their comparisons with an ideal.

Maternal mortality has long been considered a global health and development issue, and a globally accepted indicator of a nation's population health and economic development (Wilmoth et al. 2012). The United Nations prioritized maternal and child health outcomes, making them Millenium Development Goals (MDGs) and more recently, part of the Sustainable Development Goals (SDGs). Maternal mortality has improved tremendously over the past century, particularly in the last three decades, due to improvements in medical technology, better quality of life and lower birth rates. A co-occurrence of higher birth rates and lack of resources means that lower-income countries still have challenges with high rates of maternal mortality. T&T, classified as a high-income country (World Bank 2021) has put considerable effort into improving maternal outcomes (Bassaw, Seemungal, and Maharaj 2018). Therefore, the experiences and outcomes of those receiving antenatal care are a lens through which we can understand the quality of basic healthcare, and perceptions about the quality of healthcare, for a process that is both a normal aspect of healthcare, and has the potential to be high-risk, even fatal. Despite the data that support improvements in maternal mortality in T&T, a general and persistent lack of faith in the effectiveness of public health institutions remains. One of the ways women and their families overcome this is by engaging with the private sector when they can afford to do so. Those who cannot afford private care have little choice in the matter. It is perhaps a response to helplessness, an act of self-determination in the face of shocking stories of maternal death in the news, to seek private care where one might believe that payment ensures better treatment. These strategies of self-determination, borne out of inadequate or ineffective communication about maternal health outcomes in T&T, or the saliency of tragic cases reported in the news, a history of maternal mortality failures, acknowledged by the government (Neaves 2019) impact attitudes and beliefs toward public healthcare. They reinforce distrust, confusion and uncertainty. Moreover, one might argue that effective communication by institutions is undermined by historical violence against marginalized people in post-colonial space. Without a meaningful reckoning with historical violence and a radical transformation of the systems and structures that produced racial violence through colonization and slavery, no amount of effective communication can ameliorate profound institutional distrust.

The perception of inefficiency and poor maternity outcomes is linked to broader concerns about institutional reliability: how Trinbagonians perceive what is happening when they cannot easily get their needs met, accessing routine services in public institutions, the meanings they ascribe to accessing services from public institutions, and the decisions they make as a consequence. Whether there is a discrepancy between perceived state of healthcare and actual outcomes on a regional or global scale does not matter when people are distrustful of the institutions. What matters is how people experience their engagements with these institutions: whether they believe they are treated with dignity and respect; whether they believe they have some sense of control or agency while getting their needs met; whether they perceive their time as being respected. Perceived effectiveness and institutional trust are undermined if the experience of accessing services is complicated by awareness of past tragic outcomes, confusion, dismissal, uncertainty and waiting.

Waiting: The Power of Controlling Time

One aspect of interfacing with public health systems in T&T is time – waiting for information, uncertainty about when or whether you can access care or receive adequate, courteous service all remind you that you do not control time. Too often opaque explanations, miscommunication, and unnecessary steps make simple processes perplexing, rendering what should be routine transactions seemingly insurmountable challenges. In response people who are impacted the most acquaint themselves with official procedures but depend on informal or colloquial knowledge from people who have been successful before. It is no surprise, then, that wait times shape how people perceive the care and treatment they receive.

More comfortable, better and clean are words often used by those who access private care, even though they acknowledge they might get the same quality of treatment as they see the same doctors. In choosing to go private whenever they can, many hope to circumvent poor customer service – negative attitudes of nurses[5] or physician apathy, or any of the high-risk stories that litter newspapers. Women often cite lengthy waits to obtain clinic appointments, wait times to be seen, and the history of negligence leading to the death of mother or child or both as reasons to avoid public maternity care. Perhaps they can avoid the tragic experiences of Maqueder Martin, or Sherisse Williams, who hit their heads while walking to the bathroom unattended at a public hospital in T&T after having cesarean section deliveries (Wilson 2018), or Michelle Boodoo, who died of unknown causes following a cesarean section (Webb 2017). People hope that private care shields them from the uncertainty of not knowing, the arduousness of waiting, that it insulates them from tragedy. Without a doubt, horror stories elicit schemas of threat in the minds of those who must use public healthcare, and being forced to wait enacts symbolic violence through which users of the system experience submission (Auyero and Swistun 2009).

Brittney Cooper (2016) in her TED Talk: *The Racial Politics of Time*, explains time as a socio-political construct of oppression. She posits that if time were conceived as a racialized entity, it would be white. Essentially, she argues that time is constructed to function through systems of power, and in as much as structurally it is controlled by dominant, oppressive systems of power, time privileges white people. Drawing parallels between US American society with which T&T shares a history of colonialism and plantation slavery, I contend that in post-colonial T&T time operates similarly, incorporating the complexities of a multi-ethnic or plural society. The overarching ideas that govern power in this context are heavily influenced by colonialism in the Caribbean – the nature of those who held power, and those over whom they wielded power – and thus inherently contingent on Eurocentric racial (and gendered) hierarchies. Therefore, the extent to which your time is consumed trying to access services: waiting to be seen for a clinic appointment; constituent offices, the registrar's office; trying to convince a civil servant that you have the necessary documents and copies of those documents varies in proportion to the degree of whiteness (social and institutional power) you can access. Although this is rarely articulated it is repeatedly experienced, and it remains a source of frustration and confusion, limiting access and efficiency, and often dictating the pace at which people without privileged status, be it wealth, social class, and/or color, navigate everyday routine survival struggles. It appears inconsistent with capitalist objectives of efficiency that institutions would not care about time, but perhaps these constraints on the powerless reflect a more sinister indication that efficiency is a privilege mediated by the power structures implicit in the ethos of capitalism.

Speaking with friends and acquaintances and hearing conversations of those who have received reproductive care in T&T, either routine gynecological care, or maternity care through Regional Health Authority (RHA) antenatal clinics, the most often cited reason for paying for private antenatal care or a combination of private and public, was the desire to avoid excessive waiting for frequent, routine appointments, and to have a "nicer" setting, that is, a more comfortable physical environment and more friendly interactions with nursing and care staff. As such, women trade their financial resources if they can afford it, to avoid the loss of time even if they could receive the same standard[6] of care for free. Waiting, not knowing when one would be seen, not knowing why the waits are so long places a psychosocial and even economic cost on life for the marginalized, showing up not only in healthcare settings, but in public institutions where one might do business, or any space where the autonomy of the marginalized might be contested. American author, Ta-Nehisi Coates (2014), argues that "perhaps the defining feature of being drafted into the black race was the inescapable robbery of time, because the moments we spent readying the mask, or readying ourselves to accept half as much, could not be recovered" (91). Coates and Cooper's framings illustrate how the robbery of time is used as a powerful force to render routine survival struggles and the very nature of life uncertain and imperiled for Black people.

Uncertain Communication

Discourse framing uncertainty and waiting for care becomes a metaphor for inefficiency and distrust. If you can get something done quickly it must signal success. Lack of communication from physicians and other medical professionals, vague information about long waits, or whether services will be provided at a given time and defensive or curt responses from those tasked with communicating with the public fosters a sense of distrust and resentment. That miscommunication is not unique to the experience of receiving healthcare, compounds the issue. In 2021 when T&T, which had closed its ports of entry during the height of the COVID-19 pandemic, reopened its borders for international travel, the Ministry of Health of T&T issued an advisory for obtaining an international immunization card. It involved taking your local immunization card to the offices of the Chief Medical Officer in your health district. Having obtained one before the pandemic, I was familiar with the process which took no more than fifteen minutes at a health district in the capital. Conversely, a friend obtained one in a different district. Unlike my experience, she had to visit an office an hour away from her home where she and others waited for hours only to be told that they had to return the next day because the person with authority to sign the card was not available. She expressed frustration and wondered why they had not devised a more efficient and customer-friendly plan for coverage. Was it really necessary, she wondered, to inconvenience people in this manner?

I, too, have experienced confusion and long waits at a local passport office. While there I witnessed three different women leave without success, one because she misunderstood the requirements for obtaining a passport for her child. A clerk had been periodically coming out randomly (not in the order of when they had appointments or where they were standing in line) calling on the many people waiting to check their documents. She noticed that the mother, who had a 9:00 AM appointment and had waited outside in the heat for three hours with a toddler in her arms, did not have a recommender's stamp for the child. The woman, frustrated, said that having an appointment was futile. She had pondered leaving before, but she stuck it out, hoping the ordeal would soon end. After speaking to the clerk she left unsuccessfully, likely having to repeat the process if she wanted her child to obtain a passport. Another two women were unsuccessful because of issues regarding fathers who were not present for the appointment, and misunderstandings of what documentation or procedures were allowed to obtain passports for their children. Overhearing their conversation with the immigration officer, observing the officer's body language, it was evident she wanted to help. I overheard the women saying they had consulted an attorney – the father of their children was not a citizen of T&T and was out of status. These details did nothing to alter the fact that they had spent several hours at the office that day and would walk away without what they had come for.

I shared my experiences and observations with a friend who had braved the passport office a few weeks before me. She shared how, despite their appointment, her family spent five hours at the office, and that due to different (and recently revised) guidelines for renewal of her minor son's passport they almost did not get the application processed. She recounted how she was lucky the officer noticed that one of the people who signed as a reference on the application was her former teacher. My friend placed her faith in that connection. She adopted an ingratiating stance with the hope that she would be helped. The officer, who initially said they would have to come back, went to ask permission for a work around from her supervisor. My friend was relieved, but said they warned her the process could still be upended when the documents went to the head office. As we discussed our experiences and how confusing and perplexing the guidelines were, I searched, as a comparison, for the US instructions to obtain a passport. As I read them to my friend she laughed cynically, in disbelief. Of course, the United States is no standard bearer for equality or efficiency for many who are marginalized and historically oppressed. But it is understandable when examples like these are juxtaposed that people think private or foreign resources are better.

In 2016, while acquainting myself with country-level health and economic data I was surprised to learn how well-positioned T&T was compared to many other countries not only regionally, but globally in the areas of gross domestic product, health and education. To date T&T has financial wealth, economic stability, and a high standard of living relative to many other Caribbean (and Latin American) nations due to its petrochemical and manufacturing sectors (World Bank 2017). However, this does not necessarily translate to better public health systems that help the most vulnerable women. For poor women who typically rely on public services to meet basic healthcare needs, the context in which they navigate in a system that they experience as uncertain, unsafe, and unequal is one where they confront under-resourced settings, and rough treatment in the midst of signs of private and public wealth.

I have found in conversations with those who do not typically engage with country-level development data, Trinbagonians often express disbelief at our performance relative to developing nations and large countries perceived as more advanced than us. While at a meeting with stakeholders working with youth in the health sector, I noted that the T&T public health system ranks better than many places in the world, and that what our clients face is more about problems with linking to services, and how they are treated within the healthcare system, rather than a lack of treatment resources. The reaction from my colleagues, who surely knew more about the health system than I did, was skepticism. This partly is a consequence of perception: long waits, poor customer service, and news reports of tragic outcomes. They undermine the good within the system, they leave people feeling frustrated, dissatisfied, and believing that conditions in T&T are subpar.

The Consequences of Distrust

That people would express disbelief in the relative efficacy of the healthcare system is not surprising. A friend once shared that even knowing a negative story about her doctor's outcome with a previous delivery, she chose to return to her when she delivered her second child. With resignation she said, there's something in the news about almost all of them, and you might as well go with the one you know. She felt comfortable with her doctor, whom she liked so she went back to the person she knew. As to whether she ever considered using the public hospital, she said after her first experience, emphatically no, because private facilities were nicer, and the nursing staff treated you well. In any case she mused, you're meeting the same doctors. Since her employee health program reimbursed her for a large proportion of the labor and delivery costs, and she would receive partial reimbursement through the national insurance system, she considered the private hospital worthwhile. For some the ability to access private care comes with the privilege of access to private health insurance or the resources to finance private healthcare in T&T or even, at substantially higher cost, outside the country or the region. Many in T&T, at some point, communicate some degree of skepticism or distrust about public health services and, if they can afford it, use them in tandem with private healthcare or as a last resort. For instance, people may get vaccinations and minor ailments attended to in public facilities for free, but seek private care for more complicated treatments or to avoid lengthy wait times for an appointment. Whether their skepticism is founded on actual empirical data or fears nurtured by media sensationalism, panic and poor customer service, is debatable. Until I returned to live in Trinidad as an adult, I had not navigated the healthcare system of T&T. Consequently, my own perceptions until that time had been framed as a visitor, shaped by stories and encounters of a friend or relative, or a consumer of media abroad. My experiences while living and working in Trinidad provided a new perspective on how people communicate about and perceive healthcare. Drawing on the most salient examples represented in public discourse and the media, I came to understand how perceptions of maternity healthcare served as an indicator for public perceptions on healthcare in general.

Perceptions about healthcare are certainly tied to public discourse and personal experiences of other areas of bureaucracy. As such women have suspicion or fear about public healthcare whether they have utilized it or not, whether they themselves have had negative experiences. In fact, women who have experienced private services have been found to have less favorable perceptions of public care, or at least been more willing to openly state their negative perceptions (Moonesar and Vel 2012). An almost universal belief is that engaging with government agencies, obtaining a driver's license, applying for a passport or birth certificate will entail excessive paperwork, lengthy waits, confounding and needlessly complex regulations and document requirements, frustration and uncertainty. It is no surprise that these perceptions of public services match, to some extent, private sector perceptions of consumer experiences, as evidenced by T&T's ranking on the World Bank Ease of Doing Business report, where among 59 high-income countries we rank 54th and 105th among all countries surveyed (World Bank 2010). It is impossible to extricate gender, class and race hierarchies from the experiences and perception because they shape both how people experience the environment and how they are treated as a consequence of the historical contexts of the environment. Particularly for poor Black women, as Hilda Lloréns (2019) says in describing the impacts of inequalities on Afro-Puerto Ricans, their experiences are "marked by what it means to be a dispossessed person of African descent in the Americas" (n.p.).

Healthcare exists in the marketplace and as with any other "commodity"[7] access to such a vital need is influenced by other institutional and social problems that affect society. Reproductive care, standard and necessary for women to live healthy lives, is subject to those systemic forces just like any other commodity. In T&T, where citizens can access reproductive healthcare at no cost through the public health system, or via private or public health insurance programs, there is some variability in the quality, experiences and outcomes of reproductive care. Due to the country's relative wealth and its success at meeting global maternal and child health targets, it gets less attention, training and capacity-building support measures from global international health bodies such as the Pan-American Health Organization (PAHO). Typically, capacity building and monitoring and evaluation of sexual and reproductive health and maternal health have been focused largely on women living with HIV (Gómez-Suárez et al. 2019), or on women in the top ten countries with high maternal mortality rates within Latin America and the Caribbean region (WHO 2015). Also, participation in these regional and global institutional reports is voluntary for member states, with little incentive for them to provide information. Therefore, in the regional literature there is a notable lack of information on reproductive health in T&T. What is evident is that negative perceptions of maternal health outcomes are not unfounded, but they may not be as bad as they seem, or they may have improved considerably in recent years. Nevertheless, as reputational damage has been done in the past, perceived risks may take some time to catch up to any improvements in the systems. Once the perception of distrust exists, particularly when it pertains to loss of life of pregnant women and babies, it becomes difficult to reverse the impact. If women and their families are not, or believe they are not being treated with dignity and respect, they may attribute these negative experiences to the many salient negative outcomes present in the news media. As people try to make sense of delay, inefficiency, poor customer service, their frustration and confusion at interacting with institutions of service and care in routine survival struggles they draw conclusions about the larger social world.

At an individual level people make social comparisons under conditions of uncertainty. In the absence of tangible and social bases for evaluation people make subjective judgments and assessments of their abilities that depend on comparisons with others (Festinger 1957). This is particularly so when the comparisons are aspirational. If the most salient physical and social bases for evaluation are negative interactions with institutional sources of power, it is not surprising that there should be a collective response that seeks to make sense of the social environment.

So powerful is the culture of the global north in contributing to the uncertainty and distrust, the rhetoric around election fraud and demands for recounts in the Trinidad and Tobago 2021 general election (Parsanlal 2020) and recent "pro-choice" protests against masking and vaccinations (LoopTT Online 2021b) seems eerily similar to events in the United States during the 2020 election and recent public responses to public health ordinances across the United States. Nevertheless, as discussed at the onset, the United States is no stranger to social inequalities, not only in maternal health outcomes, but numerous indicators for marginalized people. Racial disparities in educational outcomes at all levels including special education (Fish 2019); and disproportionate punishment of Black children in schools (Morris 2016; Riddle and Sinclair 2019; Fadus et al. 2021) tell us an upward comparison is not a guarantee in a wealthier country if you belong to a historically marginalized group. When Black parents seek better educational options racial disparities reflect the disadvantages that lead to their crimes and the punishments they receive (Lowery 2019). Racism has been identified as a main driver of racial disparities in healthcare in the United States (Mateo and Williams 2020) and in the mortality rates from COVID-19 for essential workers (Rogers et al. 2020); even in those who is deemed in need of a psychiatric decisional capacity evaluation (Garrett et al. 2022).

Conclusion

Social environments that are contextualized by gender, racial and class hierarchies and deeply entrenched histories of violence inevitably recreate narratives of oppression for marginalized people. These systems that oppress are not unique to one country if they exist as a global, pervasive ideology of oppression based on racial capitalism. They render the marginalized – working-class people, Black people, marginalized women and anyone lacking wealth or social capital – vulnerable, persistently fighting for the certainty and security that the powerful command. For the person who has the resources to accommodate missing hours, a day or half-day's work and pay to line up at an office for service, track down ample amounts of documentation, reschedule and return if things go awry, inefficiencies are an inconvenience. They still come at the cost of time and money, but the impact is mitigated by relative wealth and a safety net. For the poor these inconveniences viscerally impact their material conditions, even at times threatening their survival. A half-day's or day's lost wages may mean unpaid school fees, unpaid utility bills, no meat on the table for this week, or even eviction. When people who are already disproportionately marginalized encounter challenges in the medical system it may deter seeking or following through with critical medical care, thus exacerbating the reciprocal impacts of illness on poverty. As such, when people cannot trust "official" information, they make calculations based on their own understanding of reality. They know government or corporate institutions may not be accountable to them, so they make decisions to rely not on official statements or data, but on their networks and themselves. These elements of the experience are important in understanding how communication failure, confusion, interpersonal turmoil, lack of information and self-determination create distrust and serve to simply wear people down. Trying to make sense of the reality of systemic oppression adds the burden of constant mental processing that marginalized people must perform to survive. Grappling with a constant state of uncertainty is incrementally debilitating, it is as if you are attempting to stream video on your computer while running too many programs in the background (Cottom 2013).

> Just as the extra processing —invisible to the naked eye—impacts the video experience, the cognitive version compromises the functioning of our most sophisticated machines: human bodies... Imagine the productivity of your laptop when all the background programs are closed. Now imagine your life when those background processes are rarely if ever activated, simply because of the social position your genetic characteristics afford you.
>
> (n.p.)

The collective history in the Americas of plantation slavery and colonialism, the commodification of healthcare, poor communication and confusion from institutions serve to create an environment that fosters uncertainty. People are unsure what information is trustworthy, and so they have little to rely on in determining their best options beyond their assessments of the landscape marked by historical violence, disservice and deception.

Notes

1 Security at most governmental agencies is operated by a government-owned security and custodial services company. It is typical for security guards to be utilized to prescreen documentation or to verify that patrons can gain access to services they seek, typically outside the health sector. Nevertheless while security guards wield a form of power, just as nurses who perform care work, they too experience various forms of oppression as workers.

2 Trinbagonian – colloquially used, an amalgamation of Trinidadian and Tobagonian, as in people of the nation of Trinidad and Tobago.

3 The climate crisis presents an impending environmental disaster beyond the type that occurred in Flammable Town, Argentina. Life-threatening impacts of the climate crisis present an existential threat to island nations around the world and continue to inequitably harm and displace the poor, particularly women. Due to increased temperatures, erratic weather patterns, flooding and worsening pollution in many regions, including the Caribbean, the climate crisis in its current state already exacerbates the burden of routine survival struggles. https://theglobalamericans.org/reports/the-caribbean-extreme-vulnerability-climate-change/. https://publications.iadb.org/en/background-paper-lac-small-island-development-states.

4 Trinidad is used here as it is in the article being referenced by Ruzdik. In my writing, I use Trinidad and Tobago to describe the nation, and Trinidad when speaking specifically about the island of Trinidad, or Tobago when speaking specifically about the island of Tobago.

5 As qualified staff nurses are better off than many of the patients they treat in the public sector. Although their salaries are not that high, particularly compared to the energy industry or nursing outside of T&T, their position in society and the

access to other opportunities places them often at a higher position than the patients they treat. Nevertheless, their status as care workers, people who possess similar marginalized identities as those for whom they care, and disproportionately women in hierarchical medical settings where patriarchal power differentials are reinforced. As Patricia Hill Collins (2002) puts it, the physically demanding and economically exploitative nature of their work sometimes inscribes them in a position of powerlessness similar to their patients.

6 I use "standard" here to mean care that meets the same benchmarks, from physicians, nurses and other medical professionals with the same technical capacity, trained in the same national institutions, qualified and licensed by the same governing bodies. Many clinicians are employed by the regional health authorities and treat patients at public hospitals, clinics and district health facilities moonlight at private facilities, or are sometimes the owners of these facilities. As such, what differs is the availability of appointments and medical equipment, the ambience of the treatment facilities and the customer service.

7 I have purposely used the term commodity here, although I have a problem with the commodification of any type of healthcare. However, this is intentional because I want to look at the impact on marginalized and disadvantaged women as they access care, and how reproductive healthcare as a commodity is accessed by those with more resources, and how it is treated by those who have more wealth and power.

References

Ali, Azard. 2019 "Saturday Woman Wins Medical Negligence Lawsuit." *Trinidad and Tobago Newsday*. Accessed 20 February, 2022. https://newsday.co.tt/2019/03/16/woman-wins-medical-negligence-lawsuit/

Allahar, Anton L. 2003. "Racing' Caribbean political culture: Afrocentrism, Black Nationalism and Fanonism." *Modern Political Culture in the Caribbean*: 21–58.

Auyero, Javier, and Débora Alejandra Swistun. 2009. *Flammable: Environmental Suffering in an Argentine Shantytown*. New York: Oxford University Press.

Baranov, D. and K. Yelvington. 2003. "Ethnicity, Race, Class, and Nationality." In *Understanding the Contemporary Caribbean* edited by Richard S. Hillman and Thomas J. D'Augustino, 209–34. London: Lynne Reinner Publishers.

Barnes, Natasha. 1997. "Face of the Nation: Race, Nationalisms, and Identities in Jamaican Beauty Pageants." In *Daughters of Caliban: Caribbean Women in the Twentieth Century*, ed. Consuelo Lopez Springfield, 285–306. Bloomington and Indianapolis: Indiana University Press.

Barriteau, V. Eudine. 2001. *The Political Economy of Gender in the Twentieth-Century Caribbean*. New York: Palgrave.

Bassaw, B., Seemungal, T., and V., Maharaj 2018. "A 36-Year Review of Maternal Deaths in a Low-Resource Country." *Obstetrics Gynecology International Journal*, 9(6): 520–4.

Beckles, Hilary. 2013. *Britain's Black Debt: Reparations for Caribbean Slavery and Native Genocide*. Kingston: University of the West Indies Press.

Blaize-LaCaille, E. Charmaine. 2018. "Laboring for Quality: Crossing the Chasm in Maternal Health Care in Trinidad and Tobago," PhD diss., (University of Stirling).

Bolland, O. Nigel. 2005. "Reconsidering Creolisation and Creole Societies." In *Contesting Freedom: Control and Resistance in the Post-Emancipation Caribbean*, eds. Gad Heuman and David Trotman, 179–96. Oxford: Macmillan Publishers.

Bourdieu, Pierre. 1991. *Language and Symbolic Power*. Cambridge: Polity Press.

Brathwaite, Edward K. 1971. *The Development of Creole Society in Jamaica, 1770–1820*. Oxford: Oxford University Press.

Brereton, Bridget. 1979. *Race Relations in Colonial Trinidad 1870–1900*. Cambridge: Cambridge University Press.

Bryant, Allison S., Worjoloh, Ayaba, Aaron B., Caughey, and A. Eugene, Washington. 2010. "Racial/ethnic disparities in obstetric outcomes and care: prevalence and determinants." *American Journal of Obstetrics and Gynecology*, 202(4): 335–43.

Coates, Ta-Nehisi. 2014. "The Case for Reparations." *The Atlantic*. Accessed 1 September, 2022 https://www.theatlantic.com/magazine/archive/2014/06/the-case-for-reparations/361631/

Collins, Patricia Hill. 1986. "Learning from the Outsider Within: The Sociological Significance of Black Feminist Thought." *Social Problems*, 33(6): S14–32.

———. 2002. *Black Feminist Thought: Knowledge, Consciousness, and The Politics of Empowerment*. New York: Routledge.

Cooper, Brittney. 2016. *The Racial Politics of Time*. TED: Ideas Worth Spreading.

Cottom, Tressie McMillan. 2013. "Whistling Vivaldi Won't Save You: The Shooting Death of Jonathan Ferrell Reminds Us of The Heavy Toll of Stereotype Threat." *Slate*. Accessed 18 March, 2022. https://slate.com/news-and-politics/2013/09/jonathan-ferrell-shooting-death-the-perils-of-stereotype-threat.html

De Silva, Radhica. 2020. "Mom Dies Walking to Hospital Bathroom After C-section." *Trinidad and Tobago Guardian*. Accessed 18 April, 2022. https://www.guardian.co.tt/news/mom-dies-walking-to-hospital-bathroom-after-csection-6.2.1258347.5dd5ef9126

Fadus, Matthew C., Emilio A. Valadez, Brittany E. Bryant, Alexis M. Garcia, Brian Neelon, Rachel L. Tomko, and Lindsay M. Squeglia. 2021. "Racial Disparities in Elementary School Disciplinary Actions: Findings from the ABCD Study." *Journal of the American Academy of Child & Adolescent Psychiatry*, 60(8): 998–1009.

Festinger, L. 1957. *A Theory of Cognitive Dissonance*. Stanford, CA: Stanford University Press.

Fish, Rachel Elizabeth. 2019. "Standing out and sorting in: Exploring the role of racial composition in racial disparities in special education." *American Educational Research Journal*, 56(6): 2573–608.

Garrett, William S., Anita Verma, Daniel Thomas, Jacob M. Appel, and Omar Mirza. 2022. "Racial Disparities in Psychiatric Decisional Capacity Consultations." *Psychiatric Services*, appips202100685. Advance online publication. Accessed 3 August, 2022. https://doi.org/10.1176/appi.ps.202100685

Glenn, Evelyn Nakano. 1992. "From Servitude to Service Work: Historical Continuities in The Racial Division of Paid Reproductive Labor." *Signs*, 18(1): 1–43.

Global Americans. 2019. "The Caribbean's Extreme Vulnerability to Climate Change: A Comprehensive Strategy to Build a Resilient, Secure and Prosperous Western Hemisphere." High Level Working Group on Inter-American Relations and Bipartisanship. Accessed 12 May, 2021 https://theglobalamericans.org/reports/the-caribbean-extreme-vulnerability-climate-change/

Global Burden of Disease Collaborative Network. 2016. *Global Burden of Disease Study 2015 (GBD 2015) Maternal Mortality 1990-2015*. Seattle, United States of America: Institute for Health Metrics and Evaluation (IHME). Accessed 22 September, 2022. https://ghdx.healthdata.org/record/ihme-data/gbd-2015-maternal-mortality-1990-2015

Gómez-Suárez, Marcela, Maeve B. Mello, Mónica Alonso Gonzalez, Massimo Ghidinelli, and Freddy Perez. 2019. "Access to sexual and reproductive health services for women living with HIV in Latin America and the Caribbean: systematic review of the literature." *Journal of the International AIDS Society*, 22(4): e25273.

Hintzen, Percy C. 2002. "The Caribbean: Race and Creole Ethnicity." In *A Companion to Racial and Ethnic Studies*, ed. David Theo Goldberg and John Solomos. Malden: Blackwell Publishers.

Hoffman, Susie, Yusuf Ransome, Jessica Adams-Skinner, Cheng-Shiun Leu, and Arpi Terzian. 2012. "HIV/AIDS surveillance data for New York City West Indian–Born Blacks: Comparisons with Other Immigrant and US-Born Groups." *American Journal of Public Health*, 102(11): 2129–34.

Hogan, Margaret C., Kyle J. Foreman, Mohsen Naghavi, Stephanie Y. Ahn, Mengru Wang, Susanna M. Makela, Alan D. Lopez, Rafael Lozano, and Christopher JL Murray. 2010. "Maternal mortality for 181 countries, 1980–2008: a systematic analysis of progress towards Millennium Development Goal 5." *The Lancet*, 375(9726): 1609–23.

Hosseini, Sarah. 2019. "Black women are facing a childbirth mortality crisis. These doulas are trying to help." *The Washington Post*. Accessed 15 April, 2021. https://www.washingtonpost.com/lifestyle/2019/02/27/black-women-are-facing-childbirth-mortality-crisis-these-doulas-are-trying-help/

Hoyert, Donna L. 2022. "Maternal Mortality Rates in the United States, 2020". *NCHS Health E-Stats*. Accessed 23 February, 2022. https://doi.org/10.15620/cdc:113967

Hyacenth, G., and Crystal Brizan. 2012. "The Case of Unsafe Abortion in Trinidad and Tobago: An NGO Perspective." *Social and Economic Studies'. Special Issue on Women's Reproductive Health and Rights in Select Caribbean Countries*, 61(3): 167–86.

Jolly, Jallicia. 2021. "A Black Doctor Died After Childbirth. 4 Years Later, Her Mother and Daughter Are Still Living With The Grief. Living the Daily Toll of This Heartache is The Aftermath of Maternal Mortality." *The Lilly, Washington Post*. Accessed 16June,2021.https://www.thelily.com/an-esteemed-black-doctor-died-after-childbirth-4-years-later-her-mother-and-daughter-are-still-living-with-the-grief/

Kassebaum, Nicholas J., Ryan M. Barber, Zulfiqar A. Bhutta, Lalit Dandona, Peter W. Gething, Simon I. Hay, Yohannes Kinfu et al. 2016. "Global, regional, and national levels of maternal mortality, 1990–2015: a systematic analysis for the Global Burden of Disease Study 2015." *The Lancet* 388(10053): 1775–812.

Khan, Rishad. 2021. WHO Head Praises Rowley on Covid Fight. *Trinidad and Tobago Guardian*. Accessed 8 April, 2021. https://www.guardian.co.tt/news/who-head-praises-rowley-on-tts-covid19-fight-6.2.1291194.99d0913fbf

Liverpool, Hollis. 2001. *Rituals of Power and Rebellion: The Carnival Tradition in Trinidad and Tobago, 1763-1962*. Jamaica: Research Associates School Times.

Lloréns, Hilda. 2019. "The Race of Disaster: Black Communities and the Crisis in Puerto Rico." *Black Perspectives. African American Intellectual Historical Society*. Accessed 8 April, 2021. https://www.aaihs.org/the-race-of-disaster-black-communities-and-the-crisis-in-puerto-rico/

LoopTT Online. 2021a "UNC to Demand Recounts in 3 Marginals: 'Don't Give Up,' saysKamla."Accessed14May,2021.https://tt.loopnews.com/content/unc-demand-recounts-3-marginals-dont-give-says-kamla

———. 2021b. "Police Charge 31 After Queen's Park Savannah Demonstration." Accessed 14 May, 2021. https://tt.loopnews.com/content/police-charge-31-after-queens-park-savannah-demonstration

Loutoo, Jada. 2017. "Woman to be Compensated After Gauze Left in Her." *Trinidad and Tobago Newsday*. Accessed 20 September, 2022. https://newsday.co.tt/2017/11/29/woman-to-be-compensated-after-gauze-left-in-her/

Lowery, Annie. 2019. "Her Only Crime Was Helping Her Kids." *The Atlantic*. Accessed 14 May, 2021 https://www.theatlantic.com/ideas/archive/2019/09/her-only-crime-was-helping-her-kid/597979/

Martin, Cedrianne J., Glennis Hyacenth, and Lynette Seebaran Suite. 2007. "Knowledge and Perception of Abortion and the Abortion Law in Trinidad and Tobago." *Reproductive Health Matters*, 15(29): 97–107. https://doi.org/10.1016/S0968-8080(07)29301-2

Marto, Ricardo, Alvarez, Lourdes, and Suarez, David. 2014. "Background Paper: LAC Small Island Development States." *Inter American Development Bank*. Accessed 4 November,2021. https://publications.iadb.org/en/background-paper-lac-small-island-development-states

Mateo, Camila M., and David R. Williams. 2020. "More Than Words: A Vision to Address Bias and Reduce Discrimination in the Health Professions Learning Environment." *Academic Medicine*, 95(12S): S169–77.

McKittrick, Katherine. 2014. "Mathematics Black Life." *The Black Scholar*, 44(2): 16–28.

McMillan, Cottom Tressie. 2013. "Whistling Vivaldi Won't Save You: The Shooting Death of Jonathan Ferrell Reminds Us of The Heavy Toll of Stereotype Threat." *Slate*. Accessed 18 March, 2022. https://slate.com/news-and-politics/2013/09/jonathan-ferrell-shooting-death-the-perils-of-stereotype-threat.html

Ministry of Health of Trinidad and Tobago. 2021. Vaccination Card Update, Ministry of Health, Government of Trinidad and Tobago. Accessed 12 February, 2022 http://www.news.gov.tt/content/vaccination-card-update#.YgF0Fu7ML-Y

Mohanty, Chandra Talpade. 2003. *Feminism Without Borders: Decolonizing Theory, Practicing Solidarity*. Durham: Duke University Press.

Moonesar, Immanuel, and Prakash Vel. 2012. "Patients' Perception on Prenatal Care Management at Trinidad & Tobago." *International Journal of Economic and Management Science*, 2(3): 63–74.

Morris, Monique W. 2016. *Pushout: The Criminalization of Black Girls in Schools*. New York: The New Press.

Namasivayam, Amrita, Donatus C. Osuorah, Rahman Syed, and Diddy Antai. 2012. "The Role of Gender Inequities in Women's Access to Reproductive Health Care: A Population-level Study of Namibia, Kenya, Nepal, and India." *International Journal of Women's Health*, 4: 351–64.

Neaves, Julian. 2019. "No Maternal Deaths This Year." *Trinidad and Tobago Newsday*. November 21, 2019. Accessed 21 September, 2022. https://newsday.co.tt/2019/11/21/no-maternal-deaths-at-mt-hope-womens-hospital-this-past-year/

Pan-American Health Organization (PAHO). 2008. Regional Strategy and Plan of Action for Neonatal Health within the Continuum of Maternal, Newborn, and Child Care. Accessed http://iris.paho.org/xmlui/handle/123456789/772

———. 2017. "Health Status of the Caribbean: Maternal Health." Washington, D.C.: PAHO/WHO. Accessed. https://www.paho.org/salud-en-las-americas-2017/?p=1599

Parsanlal, Nneka. 2020. "Kamla: 'I Am Not Conceding'" *Loop TT, Politics*. Accessed https://tt.loopnews.com/content/kamla-i-am-not-conceding

Riddle, Travis, and Stacey Sinclair. 2019. "Racial Disparities In School-based Disciplinary Actions Are Associated With County-level Rates of Racial Bias." *Proceedings of the National Academy of Sciences*, 116(17): 8255–60.

Rogers, Tiana N., Charles R. Rogers, Elizabeth Van San Webb, Lily Y. Gu, Bin Yan, and Fares Qeadan. 2020. "Racial Disparities in COVID-19 Mortality Among Essential Workers in the United States." *World Medical & Health Policy*, 12(3): 311–27.

Roopnarinesingh, S., P. Ramoutar, and Bharat Bassaw. 1991. "Maternal Mortality at Mount Hope Women's Hospital, Trinidad." *The West Indian Medical Journal*, 40(3): 139–41.

Rudzik, Alanna E. F. 2003. "Examining Health Equity through Satisfaction and Confidence of Patients in Primary Healthcare in the Republic of Trinidad and Tobago." *Journal of Health, Population and Nutrition*, 21(3): 243–50.

Sharpe, Christina. 2016. *In the Wake: On Blackness and Being*. Durham, NC and London, UK: Duke University Press.

Smith, Michael Garfield. 1984. *Culture, Race, and Class in the Commonwealth Caribbean*. Jamaica: Department of Extra-Mural Studies, University of West Indies.

Tilly, Charles. 1996. "Invisible Elbow." *Sociological Forum*, 11(4): 589–601.

Trinidad Express. 2013a. "Long, long wait at hospitals." Accessed 17 September, 2022. https://trinidadexpress.com/news/local/long-long-wait-at-hospitals/article_90d80481-27f3-57b5-a31e-378f57cdc423.html

———. 2013b. "Only One Doctor at Palo Seco Health Centre: Patients Complain." Accessed 17 September, 2022. https://trinidadexpress.com/news/local/only-one-doctor-at-palo-seco-health-centre/article_1f7a7395-b162-5d9e-9d88-1ab435c41f44.html

Trinidad and Tobago Newsday. 2017. "Cancer Patient Waits on TRHA Letter to Get Treatment. Accessed 20 September, 2022 https://newsday.co.tt/2017/11/21/cancer-patient-waits-on-trha-letter-to-get-treatment/

———. 2014. Preserving Maternity Gains. *Trinidad and Tobago Newsday*. Accessed 20 September, 2022. https://newsday.co.tt/2018/05/14/preserving-maternity-gains/

Turner, Sasha. 2017. *Contested Bodies: Pregnancy, Childrearing, and Slavery in Jamaica*. Philadelphia: University of Pennsylvania Press.

United Nations Children's Fund. 2012. *Statistics and Monitoring Section. Policy and Practice. Maternal, Newborn and Child Survival. Country Profile*. New York: UNICEF.

United Nations Development Programme. 2017. "Gender and Climate change - Overview of Linkages Between Gender and Climate Change." Accessed 9 November, 2021. https://publications.iadb.org/en/background-paper-lac-small-island-development-states

UNFPA. 2017. "Sexual and Reproductive Health Thematic Brief." Accessed 13 November, 2021 https://caribbean.unfpa.org/en/news/sexual-reproductive-health-thematic-brief

Villarosa, Linda. 2018. "Why America's Black Mothers And Babies Are In A Life-Or-Death Crisis." Accessed 4 December, 2021 https://www.nytimes.com/2018/04/11/magazine/black-mothers-babies-death-maternal-mortality.html

Webb, Yvonne. 2017. "SFGH Probes Woman's Death After C-section." *Trinidad and Tobago Newsday*. Accessed 13 November, 2021 https://newsday.co.tt/2017/10/14/sfgh-probes-womans-death-after-c-section/

Wilmoth, J. R., N. Mizoguchi, M. Z. Oestergaard, L. Say, C. D. Mathers, S. Zureick-Brown, M. Inoue, and D. Chou 2012. "A New Method for Deriving Global Estimates of Maternal Mortality." *Statistics, Politics, and Policy*, 3(2): 2151–7509.

Wilson, Sacha. 2018. "Couple Wants Probe after Stillbirth at Sando Hospital." *Trinidad and Tobago Guardian*. Accessed 12 September, 2022. https://www.guardian.co.tt/news/couple-wants-probe-after-stillbirth-at-sando-hospital-6.2.1271764.78c9921ace

World Bank. 2010. “World Bank Summary Trade Statistics.” Trinidad and Tobago Trade Summary 2010 Data *World Integrated Trade Solution (WITS)* Accessed 11 November, 2021. https://wits.worldbank.org/CountryProfile/en/Country/TTO/Year/2010/Summary

———. 2017. *World Bank Annual Report 2017*. Washington, DC: World Bank. https://doi.org/10.1596/978-1-4648-1119-7

———. 2021. *Global Economic Prospects, June 2021*. Washington, DC: World Bank. http://hdl.handle.net/10986/35647

World Health Organization. 2015. “Trends in Maternal Mortality: 1990 to 2015 Estimates by WHO, UNICEF, UNFPA, World Bank Group and the United Nations Population Division.” Accessed 16 March, 2022 https://www.who.int/reproductivehealth/publications/monitoring/maternal-mortality-2015/en/

6 Black Favela Feminism

The Struggle for Survival as a Transformative Praxis

Andreza Jorge

Introduction

This chapter is part of a larger research project which investigated the development of a community-based social project called Mulheres ao Vento (in English, "Women in/to the Wind") in the Complexo da Maré (Maré), a group of favela neighborhoods in Rio de Janeiro, Brazil, and focuses on its pedagogical and political developments. Mulheres ao Vento emerged from my experiences as a dance student at the Federal University of Rio de Janeiro and the creation of a performance art proposal based on an Orixá (spirit) from the Yorubá pantheon called Iansã or Oiá. This cosmic and pedagogical element offered a lens through which to think about the relationship between Black women, their corporeality, and dance. I developed a community and educational project from this artistic work with the Black women of Maré.

The Mulheres ao Vento project stems from my history as a community leader, combined with my artistic and academic work. After I graduated with a degree in dance, the desire to continue reflecting on women's lives transcended the walls of academia and returned to my community in the form of workshops and spontaneous conversations that culminated in structured meetings to dance and think about the Black female body and favelas from plural theoretical references within Black, decolonial, and feminist studies and theories. From 2016 onward, we held weekly training meetings on topics of interest to the project and engaged in artistic creation. Alongside Simonne Alves, a friend from college who became fundamental in the creation and idealization of this collective project that is Mulheres ao Vento, I designed the training classes and project.

Since its inception, Mulheres ao Vento presented a Black feminist methodology inspired by plural representations of Black women and definitions of "being a woman" beyond Western and Eurocentric perspectives. The project aimed at expanding society's existing references about strength, beauty, intelligence, and social leadership by utilizing references to African and Afro-diasporic cultures and ancestry. It examined remnants of this ancestry in spaces of cultural and community recovery, including religious centers of African origin, artistic cultural manifestations throughout Brazil, and other spaces where

DOI: 10.4324/9781003143550-9

Black women engage in presentations of authority and leadership based in and imbued with social and civilizational values of African origins, mainly Bantu and Yoruba traditions—traditions of the main ethnic groups of African origin present in Brazil as a result of the transatlantic slave trade.

This movement sought to include the specificities of Black, indigenous, and working-class struggles through an intersectional perspective that, according to Carla Akotirene, "aims to give theoretical and methodological instrumentality to the structural inseparability of racism, capitalism, and cisheteropatriarchy" (2018, 13). Anchored in theoretical reflections of Black feminisms which materialized during the project Mulheres ao Vento, in particular what I call a *Black favela feminist praxis*, this chapter aims to question potential fragmentations and gaps existing in the Brazilian feminist movement that systematically invisibilizes, in their agendas and discussions, the lived-theorized experiences of Black women.

I explicitly place myself in the project to contribute to the conceptualization of the specific struggle of Black and favela women in Brazil. To this end, I define and develop the term Black favela feminisms by analyzing Mulheres ao Vento from a Black and decolonial feminist perspective that does not dissociate theory from practice. Theory and practice, I contend, are inextricably linked to each other in a research project that, as was the outcome of Mulheres Ao Vento, resulted in a collective effort in their/our processes of self-recovery and identity constructions as Black and favela women. In this chapter the reflections of the women interviewed help me to not only think with Black feminist theories and the ways in which Black feminists diversified movements have historically focused on the empowerment of the Black population; but they also allow me to keep in mind that by inserting specificities in the struggle for a just society, we build a pluriversality of existence and of reading the world, shedding light on the power relations that silence, subalternize, and invisibilize.

And so this chapter develops with the following objectives: i) to deepen plural and decolonial knowledge, through showcasing the experience and the development of the project Mulheres ao Vento, proposing an Afro Diasporic epistemology with the intention of combating the epistemicide resulting from the practice of hegemonic epistemologies caused by Brazilian structural racism in the field of feminism studies; ii) to broaden the discussions on racism and its various social manifestations, with the goal of overcoming it, using the Afrocentric and ancestral knowledge found in the cultural and social praxis of Black women and favela dwellers.

Methodological Path

This chapter is based on an immersive ethnographic methodology in which I took on different roles as one of the creators of the Mulheres ao Vento project, as well as a Black woman and a resident of Maré. This project was built on and engages the different and varying points of view of those who participated. Additionally, the project engaged all types of knowledge sources, placing them all in conversation, including knowledge produced through the affective and via sensations. Taking on these different but equally important roles placed me at a "crossroads" to use the metaphor created by the theorist Glória Anzaldúa (2005, 707). In this project, the role of the researcher is not that of the so-called "hard sciences," which is utilized to pragmatically separate the subject and the object. Instead, I am explicitly engaged in activist research that remains subjective and sensitive to the socio-political or local context to which it is connected. According to Vilaça (2016), activist research is when the researcher is a horizontal partner with the participants. It collectively builds and re-signifies the interpretations of the realities to be researched. Activist research also allows for a break in the silencing, erasure, and subordination of a group (in this case Black women and faveladas) who, in doing art, produce legitimate and potent knowledge. When developing our academic and professional potential, we must therefore remain attentive to the legacy of the struggles of women who came before us and of those who are producing knowledge today, be it academic or not. In the face of inequalities and structural epistemicide, our work must remain committed to an anti-racist and anti-sexist struggle. The need to affirm activist research as a methodological path is urgent.

I spent time with the women participating in the project throughout the creative and artistic processes, leading dance workshops that were based on the narratives they told in class and sometimes in private conversations. Therefore, we built affective and everyday relationships which were instrumental in supporting my theorizations around Black favela feminisms. Based on these connections, between the lives and experiences of these women and my own experience, in this chapter, alongside the existing theoretical interventions in Black feminist studies, I analyze the accounts and narratives of four of Mulheres ao Vento participants who are also Black women living in Maré. The interviews, in general, spoke to the relationship between dance, empowerment, and representation of Black women in Maré, and helped me to think of specific paths in the Black feminist struggle based on the theory of intersectionality and the recognition of the cultural praxis of Black and favela women as bases for the construction of the concept of Black favela feminisms. This relationship with the project participants and my experience with the research conducted for my master's thesis were fundamental to building a standpoint and provided a plethora of material to theorize what Black favela feminism is and its contributions to Black feminist thought and beyond.

Producing *Afrontamentos*: Black Women and Their Struggles

Being a Black woman of the diaspora in a region currently known as "the Americas" illustrates the direct link between the lives of women and the transatlantic traffic of Africans of different cultures, who were kidnapped and enslaved by Europeans. This "mark" accompanies Black women throughout their lives. The Black woman, this Atlantic woman, as the historian Beatriz do Nascimento (1989) put it, in her existence full of challenges, brings the symbols of a body that goes against the idealized and hegemonically accepted body in society. In this sense, the Black woman's body takes on the meaning of being a body on the margin, which exceeds the limit of what is understood as positive in society. According to Kiusam Oliveira (2008), the

> female black body corresponds to a marginal body and is attentive to the social contempt imposed on black women in Brazilian society, also still stating that the human body is forged according to the characteristics of the family group, social class, religion, sexual orientation, and culture of each individual.
>
> (25)

Oliveira informs us that the dominant culture imposes a standard of how our body and phenotype should be. However, for some specific groups this standard will always be unattainable, as it refers to an ideal associated with the body and white and European phenotypes as a reference of beauty and humanity. Although the constitution of hegemonic aesthetic standards can affect society in general, people racialized as Black will always be on the sidelines and subjected to hatred and the denial of their phenotypic characteristics structurally linked to what is non-beauty and non-human, as part of structural racism. The imposition of these aesthetic-existential standards as a social rule act in a coercive and structuring way in all spheres of socialization throughout the life of every Black woman, with many everyday examples and countless situations that demonstrate these standards can never be achieved. The perversity of structural racism enhances and naturalizes these existential-aesthetic standards to the point of incessantly provoking the desire to achieve them, which can result in Black women rejecting their identity, and ties to their ancestry and traditions to feel accepted in a racist society. What is more, this identity and ancestral rejection have the capacity to continue to destroy and weaken their self-confidence and self-esteem.

And yet, despite systems that would have Black women hate themselves, Black women have also been engaged in methods of resistance and re-existence, creating and sustaining legacies which engender positive representations about themselves and their bodies and that have become increasingly known and recognized in recent years. In contemporary Brazilian society, Black women have come to recognize the importance of these legacies, for example those found in religious spaces of African origin. These spaces of cultural preservation and corporal knowledge use the subjective productions of art, dance, and spiritual traditions as methods of resistance capable of embodying/making concrete, a process through which dignity and humanity are reinstated. It is through this recognition and practice that these Black women have been helping to protect and build a nation within the Black diasporic community. Many Black women engaged in this nation-building process using the power of their bodies to re-signify and re-appropriate the marks that were imposed on them, producing alternative references of wisdom, care, and beauty in their communities.

They engage in what I refer to as *AFROntamentos* (confrontations). In the face of a society that rejects, rapes, and kills them daily, they, who over the years, have remained on the front lines of the struggle for liberation and Black transcendence. These Black women are the great Ialorixás, the "aunts of the samba schools," the great leaders in the favelas and peripheries, representatives in the churches and in the fraternities of support and take care of the community and make life every day. Their often untold stories are important.

Through the analysis of Mulheres ao Vento and its participants, I sought to tell some of these stories, using the project to create an epistemological as well as methodological space for the organization and struggle of Black women. I maintain that the work of these women in the Mulheres ao Vento project confronts ("AFROnta") models of cis-heteronormative, white, racist, phobic, and Euro-centered thinking which caricature Black Brazilian women as the other of the other. In 2010 Kilomba wrote about the other of the other,

> White women have an oscillating state, both as themselves and as the other of the white man, since they are white, but not men; black men play a role as opponents of white men, as they are possible competitors in the conquest of white women since they are men, but not white; Black women, however, are neither white nor man, and they play a role as the 'other' of the other.
>
> (124)

As the other of the other, Black Brazilian women's *AFROntamentos* are not only essential to their everyday survival but must also be acknowledged as a form of knowledge creation and intervention to the theorizing of Brazilian society and feminist studies. I argue that the *AFROntamentos*, these struggles championed by Black women throughout history, are fundamental to the movement and transformation of society and contribute to guaranteeing the lives and rights of all women.

Thinking the Plurality of Black Feminisms

Nowadays, the term feminism is so popular and viral in different spaces of society that, for many women, it has become almost an obligation to position and name oneself as a feminist. It is possible to observe references to feminism in the lyrics of popular music and on social media, and the term is frequently used by young people. Feminism, according to Ávila (2000), is understood as political practice and critical thinking; it is a movement that has the perspective of transforming gender relations, the focus of which is to fight for freedom and equality for women. For this reason, it has become a systematic and recurrent denunciation of male (particularly white male) privilege and the ways in which this privilege has historically been oppressive to women and women-identified people, as well as those who are feminized in our societies.

The European feminist movement dates, more or less, to the second half of the nineteenth century, from the organization of women claiming the right to equality, inspired by several liberation struggles around the world for the gaining of citizenship. In Brazil, it can be said that Brazilian feminism began around 1910 when Professor Leolinda Daltro founded the Republican Party for Women. One of the objectives of the Party was to reactivate the discussion about the suffrage of women in the national congress and the fight against male domination, in addition to the search for equal rights between men and women. Since then, hegemonic feminism in Brazil has developed to represent and advocate for professional women in the upper and middle classes, who are mostly white and espouse an agenda of greater access to the public sphere and of gaining equality to men. In the context of countries such as Brazil, colonized and economically organized around the system of slavery, where the confluence of racialized and patriarchal systems of oppression play an important role in the country's social and structural formation and the identity formation of its citizens, this type of feminism works to invisibilize the plight of working-class Black Brazilian women. As Sueli Carneiro (2003b) reveals to us:

> When we speak of the myth of female fragility, which historically justified the paternalistic protection of men over women, what women are we talking about? We black women are part of a contingent of women, probably in the majority, who have never recognized this myth in themselves because we have never been treated as fragile. We are part of a contingent of women who worked for centuries as slaves in the fields or on the streets, as salespeople, greengrocers, prostitutes... Women who did not understand anything when feminists said that women conquered winning rights for occupying the streets and work! We are part of a contingent of women with object identities.
>
> (2)

Carneiro's reflection demonstrates that the feminist struggle, within a context of colonization, diaspora, and slavery requires a deeper analysis of issues of race and identity construction to expand the conceptualizations and meanings of gender as an analytical category, reflecting on the superficial deconstruction of the binaries around the man-woman nomenclature. Black women cannot depend on the liberal "lean in" feminism of the white, upper-class women in Brazil, part of the small elite groups that control the means of production. While the women of the upper and middle classes struggle to occupy positions of power at work, poor Black women are demanding social and political power as they continue to hold the majority of subaltern jobs. For example, Black women engage in *AFROntas* around the minimum conditions for guaranteeing labor rights such as the maximum working hours and a minimum wage for the professions of housekeeper, nanny, and cleaning lady in which the majority of Black women are working.

The above example demonstrates that the difference in race, associated with the difference in class, becomes a component of structural oppression in the construction and history of Brazil, so that even though Brazilian women have gender in common, the starting point for their struggles could not be more different. Black women resist *(AFROnta)* the dehumanization essential to these intersecting systems of domination, and historically, their struggle for a more just society for all has been fundamental to the transformation of society and has contributed to guaranteeing the lives and rights of all women. As a favelada myself, I know all too well the struggle of Black and poor women in Brazil against a racialized class system that subordinates them/us and as such, see the urgent need for an analysis that addresses their lives.

It is important to emphasize here that, when using the concept of race, I refer to the need to problematize the way people are racialized in their daily social practices, stating that the concept of race, as it is understood today, is not biological, but is a concept loaded with ideology, because like any ideology it hides something that has not been proclaimed: the relationship of power and domination (Munanga 2004, 27). In addition, when using the concept of identity, I follow Stuart Hall, who understands identity as a place that is assumed in a union between position and context and not an essence or substance to be examined (2009, 15). Thus, the social and historical context allows the creation of cultural identities of individuals and shapes the way they are seen and socially inserted in the world. Based on Sueli Carneiro's analysis above, it is possible to think of a Black woman's identity as a historical and political subject.

Examining Black women as historical and political subjects requires a different type of theorizing and analysis than those of hegemonic Brazilian feminists, one that allows them to define themselves, to shake this image of "other of the other" and which makes visible the processes of resistance and struggle full of the inventiveness and viscerality of Black women. It requires a Black woman-centered analysis which according to Black feminist Patricia Hill Collins (1986) highlights an "insistence of black women to define themselves, to evaluate themselves" (105). According to Collins this type of analysis is necessary to "define and value the awareness of" Black women's "own self-defined point of view against images that promote self-definition in the form of an objectified "other" [as a means of] ... resisting the dehumanization essential to systems of domination" (105). This approach, we know as Black feminisms, is a vast and plural conceptual and methodological field which starts from a place of examining Black women's subordination as marked by work, racism, and the remnants of slavery. It focused on their lived experiences across time and space. Black feminisms' focus on Black women's experiences has required the field to be in constant motion, transforming, as we continue to aspire to more inclusive processes and praxis. It is a theory of plurality, and hence why I use the word "feminisms" to celebrate all the various ways of being a Black woman and the production and validation of their/our knowledge from their/our multiple experiences.

This understanding of Black feminisms highlights its historical and contemporary role as a discursive field of action both within and beyond the academy to the advancement of gender in discussions within Black radical movements and the advancement of racial discussions within feminist movements. According to Alvarez (2014), Black feminisms is responsible for the pluralization and heterogeneity of feminists project, as Black feminists analyze the intersecting social locations women occupy in the modern capitalist society, perceiving differences between roles along racial, class, and other socially constructed demarcating lines. This plurality which occurs within Black feminists fields is definitely evident in the way Black feminisms also engages the concept of decoloniality.

In the Latin American Black feminist perspective, decolonial thinking is a very powerful framework because it brings with it immediate references and connections to indigenous women and the knowledge and history of their struggles. According to Curiel,

> if we understand feminism as the whole struggle of women that is opposed to patriarchy, we would have to build their genealogy considering the history of many women in many places-times. This is for me one of the main ethical and political gestures of decolonization in feminism: taking different stories, little or almost never told.
>
> (2009, 1)

The Latin American Black feminist movement dialogues with Brazilian Black feminism and emerges from social movements and from Black women's movements organized to legitimize their various forms of knowledge production and seeking to directly influence public policies in the region. Lélia Gonzalez points out in her text "For a Latin American feminism" (1988b) that:

> For us, Amefricans in Brazil and other countries in the region – as well as for the Amerindians – the awareness of oppression occurs, first of all, by race. Class exploitation and racial discrimination combined the basic elements of the common struggle of men and women belonging to a subordinate ethnicity. The historical experience of black enslavement, for example, was terrible and painfully lived by men and women, whether children, adults or old people. And it was within the enslaved community that political and cultural forms of resistance were developed that today allow a multi-century struggle for liberation to continue. The same reflection is valid for indigenous communities. For this reason, our presence in the Ethnic Movements is quite visible; there we amefricans and amerindians are active participants and in many cases, we are protagonists.
>
> (137)

In the above excerpt Gonzalez emphasizes the contribution that Black and indigenous women—for her, Amefrican and Amerindian—bring to social movements. However, even in these ethnic spaces, racialized women still find it difficult to be seen and heard. Gonzalez continues about this silencing of Black and indigenous women:

> Our comrades from ethnic movements reproduce the sexist practices of the dominant patriarchy and try to exclude us from the movement's decision-making spaces. And it is precisely for this reason that we seek the women's movement, a feminist theory, and practice, believing there to find solidarity as important as race: a brotherhood. But what they prove are the practices of racist exclusion and domination.
>
> (1988b, 137–138)

Gonzalez' reflections indicate the absolute critical nature of attending to the fullness of the lives of Black women using an intersectional lens, not just solely on their race or their gender. It is through being silenced by both the hegemonic feminist and ethnic movements that the Brazilian Black feminist struggle gained strength in the 1980s in Brazil. This movement pledged to make the lives of Black women visible as their struggles and resistance have always been invisibilized and silenced by colonialism and the reproduction of hegemonic knowledge.

It is the responsibility of Black feminists to emphasize that, since the period of slavery, Black women have always been engaged in resistance and insurgency against patriarchy and racism. They have strategically organized themselves to survive and achieve their goals. Within a colonial context that would cause generations of harm and trauma, these women worked to overcome oppression, becoming businesswomen, artists, marketers, and have used their money to help their communities from as far back as buying freedom of many other enslaved Blacks. There are countless examples of Black Brazilian women, community leaders and stalwarts. Take for example Tia Ciata, a Black Bahian woman who moved to Rio de Janeiro and received many people in her home, becoming very influential to the history of samba in the city of Rio de Janeiro. Or, the African Black Luíza Mahin, we still know little about her history, but what we know is that she actively contributed to the Malês revolt, a revolt led by enslaved Muslims in 1835 and is a crucial figure to the history of Bahia. These and many other Black women also led Black brotherhoods within Catholic churches and spaces of Afro-Brazilian spirituality, such as Nossa Senhora da Boa Morte, in Cachoeira, Bahia, always engaging in A*FROntamentos*, building their lives in the community. Black women have always been at the forefront of insurgency processes against racial, gender, and territorial oppression, from the colonial period to the present day in favelas and peripheries. They played an important role in the struggle of Black people.

They continue to do so by producing knowledge as a form of resistance that has seldom been made visible or recognized by colonial or hegemonic powers. However, it is undeniable that they have been full participants in the construction of society. For this reason, to speak of Black feminisms is to always emphasize the importance of these narratives, these stories, and this knowledge.

Above all, it is essential to affirm the appropriation of the term feminism, created within a "Eurocentrist" perspective, and to reframe and decenter this Eurocentric view by acknowledging plural feminisms. This is both a political and theoretical shift that enables multiple meanings and fully incorporates the struggle of multiple women.

Feminismos Negro e Favelado (Black Favela Feminisms): Reflections on Feminism through Favelada Experiences

Black feminisms therefore as critical to understanding the plurality of Black women's experiences was an essential framing for the Mulheres ao Vento project mentioned earlier. For this project, I got to talk at length with the participants about their experiences as Black favela women in Brazil. These conversations offered several insights about their views on feminisms in Brazil and about navigating contemporary Brazilian society as Black women from the favelas. The women whose interviews I highlight below are clear that as Black Brazilian women from the favelas their experiences are unique to the ways in which they are marginalized by the intersections of race, class and gender. Theirs and the narratives of other women interviewed but not reflected here indicate the emergence and practice of a type of Black feminisms which is specific to their lived experiences, what I am calling Black favela feminism.

Black favela feminism is not only attentive to race and gender but also to class. According to the US Black feminist author and activist Angela Davis (2016), race, class, and gender are intertwined to create different types of oppression. Class informs the race; race informs the class. Lélia Gonzalez (1984b), similarly, points out the importance of thinking beyond race and gender by considering class relations and how the fact that she is a Black woman and a favelada becomes an additional component of oppression and dehumanization.

> But it is precisely that anonymous black woman, inhabiting the periphery, (the favelas) who suffers most tragically from the effects of the terrible white guilt. Exactly because it is she who survives on the basis of providing services, holding the family bar practically alone.
>
> (231)

According to the Brazilian Institute of Geography and Statistics (IBGE), in Brazil, the rates of poverty and extreme poverty are higher among the Black population. 32.9% of the Black and Brown population live on less than US $5.50 per day in Brazil—a value adopted by the World Bank (2018) to indicate the poverty line in medium economies, such as the Brazilian one, and the data referring to extreme poverty. In addition to the number mentioned above, 8.8% of the Black population in Brazil lives on less than US $1.90 per day. Based on these statistics, it is possible to think about a specific condition for women living in the favela, using two paradigms created to discuss the relations of knowledge production coming from favelas and peripheries, the paradigms of absence and potency (power). Regarding the absence paradigm, Silva (2012) states that:

> The Paradigm of absence doesn't recognize strategies resulting of the authentic forms of "resilience", nor does it admit forms and lifestyles delegitimized by hegemonic social, cultural, political, and aesthetic references. They are, fundamentally, social habits developed under specific conditions of life, symbolically depreciated as an integral part of the process of corporeal territorial distinction – recurrent in the urban space.
>
> (n.p.)

Also, according to Silva (2012), the potency (power) paradigm is opposed to the paradigm of the absence of discursive optics that makes visible and recognizes the inventiveness of the so-called "needy and absent" territories,

> [a]s a counterpart to the simplification of the classification of "deprived", "disadvantaged", "unprivileged", "pauperized", "marginalized", "excluded" or "needy" territories, oppose the paradigm of absence, "the inventive power" of Peripheries – translated by Potency, or by the capacity to generate practical and legitimate responses, such as those which are configured as counter-hegemonic forms of life in society. It is the recognition of the inventive power of groups marked by social inequality and stigmatized by violence (…) In other words, popular territories and their subjects must be valued for the inventiveness that contributes to full urban life, not being depreciated as expressions of absence and deprivation, among other negative representations, such as which operate as symbolic windows in the sphere to devalue stocks, reputations, and rights demands for these territories.
>
> (n.p.)

It is necessary to understand the movements and insurgencies of women's struggles within the favelas as actions connected to the larger struggle for Black women in different spheres of society. In addition, these race/class considerations of the favela provide the opportunity to rethink the hegemonic feminist struggle in Brazil, including questioning the weight of the term as an "academic" concept. These interpretations and readings of the feminisms can then perceive the power existing in other forms of struggle. This shift is exemplified by Franco (2017) when discussing Black women and faveladas:

> Black women from the peripheries, with an emphasis on favelas, are strategic representations of democratic advances and living with differences and overcoming inequalities, due to the weight of machismo and racism and the growth of xenophobic ideology. Black women, residents of the peripheries and favelas, are active in the political, cultural, and artistic settings of the city. Even though the struggle/activism they lead is

> connected to local issues and is intimately "linked" to the objective and subjective conditions of their lives in the territory, they acquire fundamental dimensions to advance their local conditions, achieving an impact throughout the city.
>
> (92)

In the above excerpt, the author Marielle Franco (2017) illustrates the importance of recognizing the ways in which faveladas and peripheral women are impacting social transformations across the city through their actions. They are intrinsic to the processes of reinventing life that try to cope with inequalities promoted by hegemonic society. Taking into account the context of the organization of the favelas in Rio de Janeiro and the varied histories of the formation of the favelas, it is possible to find the different points of convergence between them and a connection with the women's struggle for rights and recognition. In my interviews, this was a recurring point made by the women living in Maré, especially when talking about their childhood and life history. For example, Maria Felipa shares that she was

> born in the south zone [a "fancy part" of Rio de Janeiro]. I was "transferred" to Nova Holanda [in Maré] at the age of 9. We stayed here without any money, but my mom got a job. She had to stay all week at the mistress' house and took my sister because she was a baby. I stayed at the neighbor's house for a whole week and only saw my mom on the weekend.
>
> (2019)

With this reflection, Maria Felipa highlights one of the multiple forms of organization of Black women in her life. Black women have always had to develop social technologies for survival, that is, to organize and create support networks with other women to guarantee the existence of their families. They have had to do so because there is an almost total lack of support from the State and a deep stigma about Black women. If the feminist movement can be defined as an organization of women that support each other, aim for collective growth, and seek a transformation in oppressive relations of class, gender, and race, the multiple forms of resistance and inventiveness in which Black women from favelas are engaged must be fully recognized as a part of this struggle. This feminism that is born from the favelas is what I am calling Black favela feminisms.

For this reason, expanding the understanding of feminism must begin by recognizing these actions and connecting them to the global struggle in favor of women. We must move beyond the academic term "feminism" by making visible the actions and organizational practices of Black women from peripheral neighborhoods and connecting them to the global social movements for the transformation of social structures. This political and theoretical shift/move will strengthen and encourage more women to occupy spaces of leadership that can, thereby, generate personal and collective benefits. In addition, it is important to recognize and make visible what ancestry reveals to us about a Black woman, as Gonzalez (1988c) points out:

> As we know, in African societies, for the most part, from antiquity to the arrival of Islam and Jewish Christians, the place of women was not one of subordination or discrimination. From ancient Egypt to the kingdoms of the Ashanti or the Yoruba, women played roles as important as men. In many cases, even political power was shared with them.
>
> (2)

Similarly, in the favela and periphery, there is a significant representation of women in leadership roles, such as women in charge of social and solidarity actions that benefit the community, women at the head of local political leadership for improving public services, women that organize and lead associations of mothers whose children are victims of State violence, among many others. This fact goes against all the stereotypes of Black and favelada women as "mere" cooks, cleaning ladies, servants, or prostitutes (Gonzalez 1988c), stereotypes which justify the naturalization of class and racial oppression with arguments focused on blaming Black and poor people.

This extensive history of exclusion and perversity to which the Black population has been systematically subjected has a number of consequences. The fault for their life trajectory is assumed to be their own, a logic that the Black intellectual Kabenguele Munanga (1988) understands as the "perfect crime" of the "construction of racism in Brazil." To reflect on the term and the theme "feminism" and to foster this debate in the favela and peripheries is to be in connection with the growing appropriation and "popularization" of terms and the reinvention of meanings generated from an academic term. Consequently, a diversity of strands corroborates the idea that the terms need to be included and questioned so that they can take on a new and broader meaning that better approximates gender with the race and class agenda. Such a move is beneficial for the population as a whole, but mainly for a subaltern population, made up of Black people, especially Black women, favela residents and their understanding of structural racism.

For this reason, it is important to emphasize that although hegemonic feminist theories and movements (mostly white and which do not place the racial and class marker as fundamental) present gender inequality as a fundamental point of the oppression suffered by women, Black and favela feminisms, by intersecting the markers of race and class, will also include the experience of Black men in this debate on oppression. Black men also make up part of the majority of working and poor classes in Brazilian society and are also victims of structural oppression based on race and class. The concept Black favela feminisms recognizes the lives of Black men in its claims of justice for Black women understanding the racial and class markers as fundamental for the understanding and transformation of society.

In addition to the gender marker, the oppressions of race and class exert historical and structural violence on Black men, and this violence is directly connected with the lives of Black women, after all these men who are targets, for example, of genocide and incarceration en masse are children, husbands, fathers, and brothers, community members. Gender oppression and oppression based on race and class therefore need to be addressed in a cross-cutting way so that there are no false alliances with systematic policies which perpetuate the death of members of the Black population. To think and theorize about Black and favela feminisms is therefore to shape the hegemonic theories of gender oppression that neglect the racial and class experience and to commit to the inclusion of Black men and male-identified folks for the construction of a just and less oppressive society. According to Gonzalez (1979), it is worth mentioning that, "the presence of Black women (within feminist movements) has been of fundamental importance, since, understanding that the fight against racism is a priority, she does not advocate a type of feminism that would distance her from her brothers and companions" (6).

When I asked Black women from the favela about what they knew about feminism, the responses inevitably demonstrated that for them feminism had to be an intersectional movement. For example, when I asked this question to Acotirene, she replied,

> I've heard of feminism but I didn't really understand what it meant. Some women say that feminism is women's struggle against men but I don't think so. In my view, I don't see it that way; I see feminism as a kind of a struggle, a voice that women are having now to be able to shout what they want.
>
> (2019)

Then, I asked whether Acotirene thought there was a difference between the struggle of Black and white women, she replied,

> ...The need is the same but the black woman fights for racism too, because she suffers these things a lot. The white woman, so to speak, fights more for the sake of independence. ...Independent of our color, let's say that the struggle of white women is more about aesthetics. When I speak of aesthetics, it is not aesthetics of the body, it is a superficial thing. Theirs is not a fundamental fight, "we will fight for this or that" no. The black woman already has, in fact. The black woman is fighting for a victory in truth, which we do not have. Many vacancies for those who study for years are not given to black people, they are for lighter-skinned people, they are not for people of our color. There are very few black lawyers. Let's put it this way, you go to the cleaning company and you see more black people. There are people who are cleaning that have finished their studies in a good course, do you understand?
>
> (2019)

With Acotirene's responses, it is possible to perceive the contradictions of a discourse about feminism that is considered "universal" and "pluralized." At first, she seems to equate women's conditions due to the fact of being a woman, but as she goes on, she talks about differences, which make race fundamental to how this woman understands oppression and disadvantage in society. In Acotirene's testimony, there are examples of how, simply because she is a dark-skinned woman, her understanding of feminism starts from a different reality, and, therefore, it needs a different set of strategies. In speaking about feminism through her own experience, Acotirene is able to distinguish between her needs as a Black woman and those of other women in Brazilian society.

Another example of how Black favela women theorize about the specificities of their lives is seen in my interview with Eva Maria. Eva Maria developed a theorization regarding Black women's lives in the favelas based on a consciousness of the continuous colonial processes of objectification, dehumanization, and constant threat to life and naturalization of violence and brutality over their bodies. We, Black favela women, are still literally fighting for our humanity, our lives, and our daily survival against the systematic extermination and incarceration of the Black, poor, and favela populations. Eva Maria makes this clear when she reflected on this prior need to claim humanity for the Black population:

> ...Yesterday I was watching a report from a mother, a black mother, who was saying that she lost her son, complaining that the military police shot him. The military police were claiming it was a "stray bullet" in the head. Then the mother was showing the boy's work card and cap to say that if it was a stray bullet the cap had to have a hole and the woman said: On the day that there was that homicide of the boys from Costa Barros with 111 shots. I didn't let my son go with them, my son survived and he could have been one of the victims of that massacre and now my son died the same way. The only thing I want is to prove my son's innocence, he was not a criminal." And I was thinking like this: Guys, we die all the time, we die while ... *We die in our bodies, right?* They kill our bodies daily, taking away our rights, and then when we die they still want to kill our memory, put us in a marginalized place, you know? ...I think this is one of the main problems that a black woman faces and a white woman is wanting equal pay (to men), that they can get, amen. I just want to have the right to go out to work, okay?
>
> (2019) [emphasis mine]

Eva Maria's account corroborates the aforementioned perspective of understanding the Black body as something worthy of humanity. A phrase mentioned by Eva Maria and emphasized by me, "We die in our body, right?" is a very subjective phrase and takes me to such a dimension of pain, illustrating a specific type of death that to some extent differs from the "natural act of dying.". She talks about the way in which Black and poor people, who inhabit the spaces of favelas, systematically have their lives taken and their bodies affected in a structural way, that is, as part of a structure of pain and death production historically directed to these bodies, to that group of people.

Every death, understood here as loss of life, is a death of the body. However, when stating that "we (Black and favelada) die in our bodies," she refers to a greater death. Death of the body at different levels of humanity and subjectivity. The physical death of Black bodies was and continues to be naturalized, albeit with increasingly cruel ends, such as by the impact of "stray bullets." Much of what is written here was written with the sound of gunshots in the background, forged in moments of fear and hopelessness, which is part of the lived reality of being a Black woman and a favelada in Rio de Janeiro. Therefore, it is necessary to have specific strategies to face this situation. The theories and practices of Black feminisms and the strategies of confrontation help us understand the struggles that Black women over the last century waged to improve living conditions, equal opportunities, and rights over their own bodies. This is still a struggle that Black and favelada women must face daily.

This way of thinking appears in yet another account of the interview participants, Tereza de Benguela. When asked about what feminism means to her, she replies:

> It is a strength for women, regardless of whether women agree with all the items or not, but everything is related to women, and if you do not join with women to overcome these causes, who will fight? So regardless of whether you agree or not, you have to be there together because even what doesn't serve me can serve another woman.
>
> (2019)

The above excerpt exemplifies the plurality of causes feminisms must embrace and how we must all struggle together, even if the "problem" or a "threat" is not directed at your life. It is a much bigger struggle for an increasingly less asymmetrical society that does not guarantee the full life of Black women. Collins (1990) describes the need for Black women to identify ways to simultaneously pivot and stay rooted, such that we can advocate for our own causes while recognizing and validating those of others. Black favela feminisms as a counter-hegemonic epistemology and praxis therefore recognize the importance of understanding society and issues outside of one's own as a form of resistance that is inclusive and pluriversal. The history of Black Brazilian women mentioned earlier in this chapter is one of confrontation but also of building safe spaces, of politically and culturally organizing, building communities where they could express themselves and maintain their existence and escape the control of the dominant ideology. Black women's cultural and political, religious and civil, and formal and informal associations are examples of such safe places. Due to the reception, protection, and exchange which occur within these safe spaces, oppressed groups can start to identify and portray themselves, preventing self-understandings that come from the other.

Conclusion

Based on the reflections from interviews mentioned throughout this chapter, the possibilities of defining a concept that is premised on including the life experience and actions organized and led by women from peripheral and systematically invisible communities are expanded. Creating a term like Black favela feminisms opens the door to problematizations about the involvement of these women who are "socially excluded" and who, in many cases, are on the fringes of what is considered "intellectualized" by canonical, racist, and imperialist structures of knowledge production.

When considering and making visible the agentic actions of these women to maintain their lives and their communities, their great power as leaders and producers of critical thinking becomes visible. With this recognition, we can consider their actions and productions in reformulating theories within feminist studies, reinventing concepts, and continuing to fight for the emancipation of all women and for social equality of race, gender, and class.

Therefore, producing knowledge involved in validating the lived experience of Black and favela women as indispensable agents for the epistemological constructions of the feminist movement is, above all, a visibility conference for those who have always been at the forefront of community organization processes, popular resistance for the guarantee of rights and survival. It is the experience of these women who continue to contribute with reflections that reallocate the linear and universalizing thinking of those who experience social and colonial oppression. The Mulheres ao Vento project, as an artistic and subjective demonstration of the creative-reflective power of Black women from the favelas, is configured as an intellectual landmark of rescue practices, resistance, and the historical-cultural memory of Black women. This is an example that moves and enhances discussions about being a woman, being Black, and being from a favela when placed in dialogue with Black feminist studies. This experience embodies the struggle for survival as a transforming and potent praxis.

Bibliography

Alvarez, Sonia E. 2014. "Para além da sociedade civil: reflexões sobre o campo feminista." In: *Dossiê o Gênero da Politica: Feminismos, Estado e Eleições*. Cadernos Pagu 43 July–December 2014. pp. 13–56. São Paulo. SP.

Alves, Heliana Castro. 2016. *Eu não sou o milho que me soca no pilão: jongo e memória pós colonial na comunidade quilombola Machadinha-Quissamã*. Rio de Janeiro.

Akotirene, Carla. 2018. *O que é interseccionalidade?*. Belo Horizonte - MG: Letramento: Justificando.

Anzaldúa, Gloria. 1981. "Speaking in Tongues: A Letter to Third World Women Writers." In *This Bridge Called My Back: Writings By Radical Women of Color*, eds. Cherríe Moraga and Gloria Anzaldúa, 165–74. New York: Kitchen Table Press.

———. 2005. "La Conciencia de la Mestiza: Rumo a Uma Nova Consciência". *Revista Estudos Feministas*, 13(3): 704–19.

Ávila, Maria B. 2000. "Feminismo e Sujeito Político." *Revista Proposta.* N° 84-85, FASE, ano 29. São Paulo, pp. 6–11.

Carneiro, Sueli. 2003a. "Mulheres em Movimento." *Estudos Avançados*, 17(49): 117–32.

———. 2003b. "Enegrecer o Feminismo: Asituação da Mulher Negra na América Latina a Partir de Uma Perspectiva de Gênero". In *ASHOKA Empreendimentos Sociais*, TAKANO Cidadania (Orgs.). Racismos contemporâneos, 49–58. Rio de Janeiro: Takano Editora.

Collins, Patricia Hill. 1986. "Learning from the outsider within: The sociological significance of black feminist thought". *Social Problems*, 33(6): S14–32. Oxford University Press. Special Theory Issue October–December 1986.

———. 1990. "Black Feminist Thought in the Matrix of Domination." In *Black Feminist Thought: Knowledge, Consciousness, and the Politics of Empowerment*, ed. P. H. Collins, 221–38. Boston: Unwin Hyman. ISBN-10:0044451377.

———. 2002. *Black Feminist Thought: Knowledge, and The Politics of Empowerment.* 2ª ed. New York and London: Routledge.

———. 2004. *Black Sexual Politics: African Americans, Gender and the New Racism.* New York: Routledge.

Curiel, Ochy. 2002. "Identidades Esencialistas o Construccion de Identidades Politicas: El dilema de las feministas Negras. Revistas Otras Miradas, Grupo de investigación en Gênero y Sexualidad." *GIGESEX*, 2(2): 96–113.

Davis, Angela. 2016. "Mulheres, Raça e Classe." *Trad. de Heci Regina Candiani*, 244. São Paulo: Boitempo Editorial.

Franco, Marielle. 2017. "A Emergência da Vida para Superar o Anestesiamento Social Frente à Retirada de Direitos: O Momento Pós-golpe Pelo Olhar de uma Feminista Negra e Favelada." In *Tem Saída? Ensaios Críticos sobre o Brasil*, eds. Winnie Buenno, Joana Burigo, Rosana Pinheiro-Machado, and Esther Solano. Porto Alegre: Editora Zouk.

Gonzalez, Lélia. 1979. "O Papel da Mulher Negra na Sociedade Brasileira: Uma Abordagem Política Econômica." *Spring Symposium the Political Economy of the Black World.* Los Angeles: Center for Afro-American Studies, UCLA, 10–12 de maio, 1979, (mimeo).

———. 1988a. "A categoria Político-cultural de Amefricanidade." *Tempo Brasileiro, Rio de Janeiro*, 92/93: 69–82.

———. 1988b. "Por um Feminismo Afrolatinoamericano." *Revista Isis Internacional. Santiago*, 9: 133–41.

———. 1988c. "A importância da organização da mulher negra no processo de transformação social." *Raça e Classe*, Brasília, ano 2, n. 5: 1–2, nov./dez.

———. 1984a. "O Papel da Mulher Negra na Sociedade Brasileira: Uma Abordagem Política Econômica. Spring Symposium the Political Economy of the Black World. Los Angeles: Center for Afro-American Studies, UCLA, 10-12 de maio, 1979, (mimeo).

———. 1984b. "Racismo e Sexismo Na Cultura Brasileira. In *Revista Ciências Sociais Hoje*, ed. Cortez Anpocs. 223–44. São Paulo/SP: Ed. Cortez.

Hall, Stuart. 2009. *Da Diáspora: Identidades e Mediações Culturais.* Belo Horizonte: Editora UFMG.

Hooks, Bell. 1995. "Intelectuais Negras." *Revista Estudos Feministas*, 3(2): 464–78.

IBGE – Intituto Brasileiro de Geografia e Estátistica. 2022. Desigualdades Sociais por Cor ou Raça no Brasil - *Estudos e Pesquisas Informação Demográfica e Socioeconômica*, n.48, 2ª edição. Rio de Janeiro: IBGE.

Kilomba, Grada. 2019. "Memórias da plantação: Episódios de racismo cotidiano". In *Trad. Jess Oliveira*, ed. Rio de Janeiro, 249. Cobogó.

Munanga, Kabengele. 1988. Construção da identidade negra: diversidade de contextos e problemas ideologicos. In *Religião, Politica, Identidade*. orgs. Josildeth Gomes Consorte, Marcia Regina da Costa. Série Cadernos PUC, 143–46. São Paulo: Educ. Available from: https://biblio.fflch.usp.br/Munanga_K_ConstrucaoDaIdentidadeNegraDiversidadeDeContextosEProblemasIdeologicos.pdf

———. 2004. Uma abordagem conceitual das noções de raça, racismo, identidade e etnia. In *Programa de Educação sobre o Negro na Sociedade Brasileira*. Niterói: Editora da Universidade Federal Fluminense/EDUFF. pp. 17–34. Niteroi/RJ. ISBN85-228-0380-3.

Nascimento, Beatriz and Gerber, Raquel (Dir.). 1989. Ori. São Paulo, Angra Filmes. 90 min.1989. Movie Orí.

Oliveira, Kiusam Regina. 2008. *Candomblé de Ketu e Educação: Estratégias Para o Empoderamento da Mulher Negra*. São Paulo, USP-São Paulo University.

Silva, Jailson. 2012. "Para uma Pedagogia da Convivência na Cidade." In *O Novo Carioca*, ed. Rio de Janeiro, 211–17. Morula.

Vilaça, Aline Serdezello Neves, "Diz!! Jazz é dança: Estética Afrodiasporica, pesquisa ativista e observação cênico coreográfica". Dissertation PPRER/Program in CEFET - Rio de Janeiro, 2016. Available: https://dippg.cefet-rj.br/pprer/attachments/article/81/76_Aline%20Serzedello%20Neves%20Villa%C3%A7a.pdf

Section III

Black Feminist Activism: A Worldmaking Praxis of Care

7 Subversive Knowledges and Praxes of Black Immigrants in the United States

Reflections from a Scholar-Advocate

Nana Afua Yeboaa Brantuo

Introduction

Lucille Clifton's (1983) poetics in *Won't You Celebrate With Me* functions as an epistemological and methodological intervention for me as a researcher, scholar, and advocate. Born the last girl child of working-class, West African immigrants in the heart of US political decision-making, I navigated barriers of systemic oppression with little guidance as did my peers and the larger familial and community networks they belonged to. While surviving and navigating, the necessity of uncovering, recovering, and recording our past and present lives became deeply personal and political. Our societal positionalities, across nation-states, exist at the intersection of anti-Blackness, misogynoir (Bailey and Trudy 2018), and cisheteropatriarchy (McLean 2021) across the time/space that is Babylon—the "present Liberaldemocratic nation state and Western world system" (Wynter 2007). Despite a world committed to our systematic marginalization, exclusion, erasure, and oppression, "we, as Black people, have been able to shape our lives and the lives of others" and "looking forward, [have] invested in futures we can't quite grasp yet…wishing, hoping, aiming at everything that has been deemed impossible" (Olufemi 2021, 1). The lives of Black immigrants in the United States, specifically those of advocates and organizers committed to community care and liberation, tell transnational narratives of resistance and decoloniality in practice, unsettling and reimagining the "dismantling of several layers of complex and entrenched colonial [and settler-colonial] structures, ideologies, narratives, identities and practices that pervade every aspect of our lives" (Tamale 2020, 20). Through their mobilities, space-making, and networks, Black immigrants navigate and subvert anti-Black state-sanctioned surveillance, policing, and violence while celebrating each day that the state failed in its efforts to remove us, both physically and from the historical and contemporary record.

Interweaving personal stories and reflections with cultural and historic artifacts and scholarly works, in this chapter, I will immerse myself within and commit myself to the subversive knowledges and praxes of Black immigrants by honoring and centering the cultural, intellectual, and political traditions and productions of Black immigrant advocates' past and present.

DOI: 10.4324/9781003143550-11

Personal Geographies

Tiffany Lethabo King's (2019) *The Black Shoals* offers a necessary and thoughtful theorization of Black Studies' engagement with Native Studies and thought. The shoal is a metaphor for meeting space between two peoples, fields, and intellectual traditions as well as necessary space for reflection on the ways in which Black people "share the hemisphere with Indigenous people also experiencing the day-to-day terror of conquest" (x). Black immigrant histories and geographies exist across terrains of indigenous land dispossession and colonization, throughout the Western Hemisphere and the African continent. Forts, towns, and parishes across our lands bear the names of English kings and queens whose charters and treaties transformed and entangled our fates.

I was born and raised on Piscataway land, the child of Ghanaian and Sierra Leonean immigrants, in a small unincorporated area named Chillum. "Chillum" or Chillum Castle Manor (Riggs 1946) was established by a land patent in 1763 and owned by the Digges family. The Digges were descendants of a founder of the Virginia Company—"human rubbish from Europe, who used enslaved but noble and exalted human beings from Africa…to satisfy their desire for wealth and power" (Kincaid 1988, 80). This occurred after near decimation of the Piscataway, once a powerful and complex polity (Strickland, Busby, and King 2015) similar to the Susquehannock and the Patawomeck. They established diplomatic relations and treaties with the infamous founding fathers of the United States—who neither honored nor respected the lands, histories, and cultures of indigenous peoples across the Americas and Africa. The low-income apartment complex that I, our African American and Black immigrant neighbors called home, was erected on land that holds the complex history of the United States' inception and myth of *freedom and justice for all.* At least one dozen Black folx across generations were enslaved on the manor, mirroring the feudality and brutality of British, and later American, society and governance.

The names of streets, buildings, and bodies of water in my community, such as Powhatan and Susquehanna, marked projects of indigenous removal and dispossession. Similarly, the preservation of manors and mansions of colonists and early statesmen marks projects of reduction and invisibilization of enslaved African lives and labor. I can recall learning more in school about the arrival of John Smith in Maryland in 1608 and the subsequent colonization of Piscataway land by charter in 1632 than learning from histories and artifacts of the Piscataway and other indigenous tribes. I can recall a field trip to Riverside Plantation in middle school, a site of "suffering and sorrow…[and] darkness and torment that the sin of slavery had caused" (Plummer 1997, 3). These lands and the Black and indigenous lives lived on them are all entangled with my own and the lives of Black immigrants to the United States, past, present, and future. Laws, policies, practices, and institutions of the United States' settler-colonial government apparatus have always demonstrated how "the racialization of space is the organizing principle through which unequal and uneven development takes place, rather than the results of this development" (Summers 2019, 13). Countries of origin for Black immigrants to the United States and our geographic dispersion have always been dependent (in part) upon the level of the United States' (and European) involvement in Western hemispheric and global affairs. From British colonization and slavery across West Africa and the Americas, to present day US imperialism, conditions necessary for the systematicization and institutionalization of cisheteropatriarchal, white supremacist, imperialist governance (locally and globally) have been developed and implemented within and across the borders of the Washington, DC, metropolitan area and our migration stories collectively are necessary for reading, untangling, and revision of history.

The remainder of this chapter attempts to bring these stories together in a retelling of US history by centering the presence and significance of Black immigrants. From scholars to activists cited, and poems to song lyrics referenced, this chapter is heavily statured with Black immigrant cultural artifacts and intellectual and artistic production—from the earliest of us to arrive to those of us living and navigating through the COVID-19 pandemic.

Uncovering Early Black Immigrant Histories and Geographies

Refugees and Laborers: The First among Us

Black immigrants' histories and geographies surround us. Black migrations to the United States, however, have been handled without care by immigration researchers and historians who often silence, reduce, and remove Black lives in acts of methodological and historiographic violence. Prominent African American sociologist Ira De Augustine Reid (1938) wrote extensively on these erasures while capturing the complexities of the growing, Black immigrant community during the early twentieth century, sharing that between the years of 1899 and 1936,

> They come from the West Indies, Central America, South America, Africa, and the Azorean Island...with their diverse customs, traditions, institutions and ideas of homeland...modifying their own culture to conform to the status accorded Negroes in the United States... [and] in turn modifying the culture of the American Negro in these communities and in the country as a whole.
>
> (411)

Engaging with demographic information, research literature, government publications, primary documents, and material artifacts from as early as the eighteenth century further unveils an emergent, diverse community absorbed within and present throughout Black America as well as mobilities and migrations influenced by the United States' hemispheric and global political agenda.

Black people from the Caribbean are the earliest of Black immigrants to migrate to and establish communities in the United States. Dodson and Diouf (2004) note that "distinguished Caribbean migrants populate the annals" (158) of early Black America. Refugees fleeing Saint Domingue during the Haitian Revolution arrived in the United States in the thousands during the late 1700s and early 1800s (Lundy 2006), though their entry was met with skepticism and fear of Black uprising by US politicians. Thomas Jefferson, in his correspondence to several prominent statesmen of the time, described Black migrants from Haiti as "the cannibals of the terrible republic" (Jefferson 1799) and was certain that

> all the West India islands will remain in the hands of the people of colour, and a total expulsion of the whites sooner or later take place. It is high time we should foresee the bloody scenes which our children certainly, and possibly ourselves (South of Patowmac) have to wade through, and try to avert them.
>
> (1793)

Black Haitian migrants, whether free or enslaved, entered into, and were the first to navigate, the racialization of the US immigration apparatus—a year following the enactment of the 1790 Nationality Act, which restricted citizenship to free, white persons. Black folx, in the wake of the Haitian Revolution and emancipation across the Americas and Cape Verde Islands, and Black migrants, in particular, navigated the reconfiguration of what Cedric Robinson (2005) conceptualized as racial capitalism. He argued that

> The formal endings of slave systems of production in the nineteenth century marked the beginnings of a profound reorganization of the capitalist world system. In Europe, Africa, Asia, and the Americas, through the deepening penetrations of monopoly capitalism and the impositions of hegemonic colonialisms, slaves were displaced as a source of cheap labor power by peasants and migrant laborers.
>
> (311)

Black Bahamian and Black Cuban economic migrants to Florida would follow Haitian refugees, with early migrations beginning in the 1800s to Tampa and Miami (Dunn 1997). Migrating out of the Caribbean following severe regional economic decline and depression, Bahamians and Cubans both navigated, shaped, and complexified Jim Crow Florida alongside and often in community with African Americans. Toward the latter end of the 1800s, the abolition of slavery in Cape Verde in 1878 commenced the earliest voluntary migrations from the African continent to the United States. Young Cape Verdean men set across the Atlantic to New England, leaving behind "poverty, drought, famine, and Portuguese colonialism" (Lima-Neves 2015, 61) on American whaling ships en route to Massachusetts as laborers for captains who "paid [them] lower wages than their American counterparts" (Halter 2008, 35).

Berlin, Brussels, and Beyond: The Origins of International Diplomacy and Early Twentieth-Century Migrations

The transition into the twentieth century was marked by the emergence of the United States as a dominant actor in a larger global system and terrain of political, economic, and militaristic power. The former colony earned its seat at the imperial table by way of genocide, slavery, and dispossession and would move to the head of the table over time. Focused solely on securing "unimpeded traffic of the Congo Valley and the west coast of Africa" (US Government Printing Office 1886, 318), the United States joined Western Europe in deliberations and agreed to resolutions at the 1884–1885 Berlin Conference and 1889–1890 Brussels Anti-Slavery Conference that would catalyze and necessitate outward migration from Africa indefinitely. Signatory states divided the African continent among themselves and reserved the right to dispose and grant concessions for land and natural resource development, while continuing to "watch over the preservation of the native populations and to supervise the improvement of the conditions of their material well-being" (Allied and Associated Powers 1919) through charity/philanthropic, scientific, religious, and governmental institutions. The abolition of slavery was merely a byproduct of a larger "move into the imperialist epoch, and to colonise and further underdevelop Africa" (Robinson 2005, 161), resulting in political and economic conditions tethered to and dependent on the Global North.

In the case of US-Western Hemispheric relations, US statesmen eyed and encroached on the lands and sovereignty of Caribbean and Latin American nations. The Treaty of Peace between the United States and Spain of 1898, the Hay–Pauncefote Treaty of 1901, the Hay–Herrán Treaty of 1903, the Hay–Bunau–Varilla Treaty of 1903, and the Treaty of the Danish West Indies of 1918 concretized US imperialistic presence and rule throughout the circum-Caribbean region. Together, the treaties transformed coloniality in the region with various forms and articulations of US imperial rule in Cuba, Puerto Rico, and the Virgin Islands and extended US control over Panamanian land and waters for the purpose of constructing the Panama Canal. President Theodore Roosevelt, in his 1904 Annual Message to Congress, expanded hemispheric policy agenda setting, better known as the Roosevelt Corollary to the Monroe Doctrine. He insists that,

> It is not true that the United States feels any land hunger or entertains any projects as regards the other nations of the Western Hemisphere save such as are for their welfare. All that this country desires is to see the neighboring countries stable, orderly, and prosperous…In asserting the Monroe Doctrine, in taking such steps as we have taken in regard to Cuba, Venezuela, and Panama, and in endeavoring to circumscribe the theater of war in the Far East, and to secure the open door in China, we have acted in our own interest as well as in the interest of humanity at large.

However, between 1898 and 1934, the United States invaded and occupied Panama, Cuba, the Dominican Republic, Nicaragua, Haiti, and Honduras in the interest of banking (Hudson 2017), bananas (Martin 2018), and military expansion and control over Western hemispheric waters (Vine 2020), all of which would not be possible without Black migration, labor exploitation, and indigenous land removal and dispossession.

The emergence of intergovernmental organizations followed shortly after. The intentions and models set in place in Berlin and Brussels guided and shaped the establishment of the League of Nations in 1919; President Woodrow Wilson was its lead architect. Following the conclusion of the First World War in 1918, in which Black folx were mobilized and sacrificed in the interest of European ethno-nationalistic warfare, the League expressed in its Covenant that *other peoples*, specifically Africans and Pacific Islanders, were in need of administration by signatory countries. The Covenant also ensured equal access and opportunities for trade and commerce across the African continent by member nations of the League. The League would later evolve into the United Nations in 1945 following the Second World War and in tandem with the founding of the World Bank Group in 1944 and the International Monetary Fund in 1945. Barriteau (2001) argues that

> The 1945 Bretton Woods meetings that created the International Monetary Fund (IMF) and the International Bank for Reconstruction and Development (IBRD or World Bank) were informed by an intellectual and ideological climate dominated by the United States. This dominance transcended military superiority to include an intellectual hegemony that has been and continues to be played out in scholarship, research, and policy-making with respect to developing countries.
>
> (143)

These reconfigurations of imperial political and economic policy and governance were based on longstanding, global capitalistic commitments to displacement, dispossession, and labor exploitation in service of the Global North. For Black immigrants, settlement across nation-states—by way of treaties, agreements, and bilateral relations—necessitated "forming diasporic subjectivities through movement across multiple Souths, and through reliance on each other in navigating these spaces" (Doig-Acuña 2020, 16). Their worldmaking, equally global and longstanding, is an amalgamation of people, enclaves and communities, nations, and movements (physical and social) that have shaped history for centuries.

Origins of Black Immigrant Worldmaking, Activism, and Advocacy

The worldmaking projects of Black immigrants in the United States reaches far back, with churches and historically Black colleges and universities (HBCUs) serving as key sites for historiographic recovery. In spite of the complexities of educational attainment following the emancipation period (Span and Sanya 2019), Black students from across Africa, the Caribbean, and Latin America traveled to the United States, acquired skills, and built networks that existed in and had influence beyond their college and university campuses. Annual convention journals from The National Baptist Association from the late 1800s and early 1900s, as well as early catalogs and student newspapers from HBCUs such as Spelman (Spelman Messenger 1915) and Hampton (The Institute 1912) captured Black international students' experiences and contributions locally and globally. Black migrants such as Mother Mary Lange, founder of the Baltimore, Maryland Oblate School for Colored Girls in 1828 (Morrow 2002), and Dr. Sophia B. Jones, founder of Spelman College's Department of Nursing in 1885 (Reid-Maroney 2004), transformed the educational trajectories of students during and beyond their lifetimes.

Organizing and advocacy in the interest of Black folx and communities, regardless of citizenship status, has also been deeply embedded within Black immigrant worldmaking projects. Between 1919 and 1927, for example, the Universal Negro Improvement Association and African Communities League (UNIA), which was founded in Jamaica by Marcus Garvey, was headquartered in Harlem. At its height, the organization's vision and mission resonated deeply with working class, Black immigrants, and African Americans alike. Among some of the UNIA's top activists were immigrants such as Louise Langdon Norton Little (Collins 2020), Laura Adorkor Kofi (Newman 2014), and Carlos A. Cooks (Cooks 1992). During the same time period, Madame Stephanie St. Clair's French Legal Society was welcoming Black immigrants to New York City. St. Claire was both subjected to and subverted criminalization and policing as one of Harlem's top policy bankers (Harris 2008). Likewise, Claudia Jones' activist origins would take root in the US Communist Party during the campaign to free the Scottsboro boys, and labor organizer Maida Kemp's work with the International Ladies' Garment Workers' Union would begin.

The Harlem Renaissance (and all Black cultural and literary movements beyond) also has African, Caribbean, and Latin American imprints. Claude McKay's famous 1919 poem, *If We Must Die*, is one of the most famed sonnets of the period, arguably marking the beginning of the Harlem Renaissance and a call to increased race and class consciousness. The collected texts and artifacts of historian, writer, and archivist Arturo Schomburg were acquired by the New York Public Library in 1926 and are now housed in the Schomburg Center for Research in Black Culture. Andy Razaf's *Ain't Misbehavin'* (1929) and *Honeysuckle Rose* (1929) are both widely recognized jazz standards, securing his place in the Songwriters Hall of Fame.

The political, cultural, and intellectual productions of these (and other) members of the Black immigrant community captured the complexities of diasporic subjectivities and, in their critique of imperialism and white supremacy, were met with surveillance and policing. The federal government spied extensively on Black immigrant activists, as is captured in the declassified FBI files of Marcus Garvey, Claude McKay, Andy Razaf, and Claudia Jones. Garvey and Jones's activism both resulted in their deportation. On her deportation, Johnson (1985) who is of Trinidadian origin shared

> I was deported from the USA because, as a Black woman Communist of West Indian descent, I was a thorn in their side in my opposition to Jim Crow racist discrimination against 16 million Black Americans in the United States in my work for redress of these grievances for unity of Black and white workers, for women's rights, and my general political activity urging the American people to help by their struggles to change the present foreign and domestic policy of the United States.
>
> (n.p.)

Jones' analysis of the entanglements and intersections with the US imperial project greatly shaped her activism and necessitated her removal by the state for her clear opposition to the evolving criminalization apparatus of the United States government and the oppressive conditions it sustained and exacerbated locally and globally.

Under Pressure from Washington: Black Migration and Advocacy in Response to US Global Dominance Apparatus

Global Monetary Coercion and Contemporary Black Migrations

The World Bank and the International Monetary Fund, the United States' legislative, executive, and judicial branches (and their respective departments and subsidiaries), and the Federal Bureau of Investigation and Central Intelligence Agency concretized the political, economic, and militaristic apparatus of US global dominance. Existing neatly within Washington, DC, with no agency being no more than four miles away from the other, their buildings and conference rooms were named in memory of past statesmen who drafted and enacted policies and legislation that dictated our migratory pathways long before our entries.

Since the Second World War, capitalistic land, labor, and resource needs have fueled the apparatus of US global dominance. In 1943, as a result of an intergovernmental agreement between Britain and the United States federal government, the British West Indies Temporary Labor Program made way for Black agricultural migrants from Jamaica, the Bahamas, Barbados, and British Honduras to labor in Southern Floridian sugar cane fields (Vialet 1978). They would replace generations of African American laborers (pre- and post-emancipation period) for generations, from pre- to post-colonial Jamaica, laying the blueprint for the H-2 Worker program, which continues to depend on Jamaican labor on sugar, cherry, and apple fields across the United States. The death of Earl Edwards (Ferriss 2020), a Jamaican H-2 worker who contracted and died as a result of COVID-19 while working on Gebbers Farms in

Washington state, connects to decades of exploitative labor agreements, which have exploited and prematurely ended Black migrant lives. Dub poet Allan Hope also known as Mutabaruka captures workers' experiences in his 1989 song, *H2 Worka* (Dedicated to the Farm Workers of Jamaica).

The establishment of the World Bank in 1944 and the International Monetary Fund in 1945 to rebuild the economies of Europe would ensure the indefinite economic co-dependency of African and Caribbean colonies turned independent nation-states. In need of capital for nation-building projects following centuries of colonial rule, the World Bank's and IMF's structural adjustment programs resulted in "the abolition or liberalization of foreign exchange and import controls, the devaluation of the currency, and domestic anti-inflationary policy specifically targeted at government spending" (Barriteau 1996, 144) across African, Caribbean, and Latin American economies. The 1970s and 1980s, in particular, proved devastating, across Black nations and their diasporas in the United States. While Shirley Chisholm (also known as Pepperpot) critiqued the Reagan administration in Congress for the inherent anti-Black racism of the immigration system overall and refugee policies specifically (Chisolm 1982), deejays across borders captured the global climate. As Caribbean Diasporan deejays and hip-hop architects (Perry 2004) utilized the newly emerged genre to speak to the social conditions of urban Black America, Jamaican deejay William Maragh, also known as Super Cat captured the global state of affairs in his 1985 single *Under Pressure*.

Mass Surveillance and Criminalization of Black Immigrants and Black Immigrant Communities

Black immigrants who resettled in the United States during the Cold War did so under the pressure of macro-level fiscal and monetary policy and external lending that "contributed to the further impoverishment and marginalization of local populations, while increasing economic inequality" (SAPRIN 2002, 173). Moreover, the Federal Bureau of Investigation, the Central Intelligence Agency, and the Departments of State and Defense worked (across time) to undermine and disrupt political infrastructures and movements across borders. Declassified intelligence gathered on African, Caribbean, and Latin American international students and movement leaders from the 1960s to the 1980s (United States Government Central Intelligence Agency 1964; United States Government Central Intelligence Agency 1970; United States Government Central Intelligence Agency 1980) capture the anxieties and fears of the United States government with respect to the transnational reach and impact of Black liberation and solidarity movements. The 1980s and 1990s would bring on political interventions, invasions, and occupations of Panama, Haiti, and Grenada alongside immigration reform that "form[ed] the basis of a shifting rhetoric that conflated improper entry with unauthorized presence while targeting and criminalizing immigrants of specific racial, ethnic, national, and religious backgrounds" (Brantuo 2017). Clinton-era law that encouraged toughness and rigidity (Clinton 1996), as opposed to reparations and social welfare, was deeply rooted in anti-Black racism and exacerbated "the vulnerabilities of Black immigrants, and, moreover, the reduced visibility of this diverse population within immigration and criminal reform advocacy impede[d] efforts to advance meaningful immigration and criminal reform" (Palmer 2017, 103). Later, compounded by Bush-era immigration law, the Department of Homeland Security and its subsidiaries were established in 2001. With Immigration and Customs Enforcement (ICE), Black immigrant communities across the country watched as Washington, DC, created inhumane and unjust policies that have resulted in increased policing and surveillance, criminalization, deportation, and separation of families, loved ones, and communities (Hilal 2022).

Contemporary Advocacy for and toward Black Immigrant Futures

The disappointments of the Obama administration, the nightmare that was the Trump Administration, and the underwhelming first year of the Biden administration have been met with fierce resistance from organizations such as the UndocuBlack Network, the Black LGBTQIA+ Migrant Project, African Communities Together, Haitian Bridge Alliance, and the Black Alliance for Just Immigration. Together, in their commitment to reparative justice and policy, they have pushed forward analysis that takes into consideration "the legacy of racial slavery and post-slavery colonization… [and] long-lasting and multi-layered negative results such as mass exodus, displacement, debt accumulation, and increasing health disparities" (Baldwin and Mortley 2020, 49–50). These organizations build on legacies of and push forward necessary advocacy in service of Black immigrants, migrants, and refugees across the United States.

Conclusion

The histories and narratives born out of Black mobilities and migrations across time and space have reimagined and reconstructed local, national, and global socio-political terrains. Despite the origins and modern functions of racial categorization and racialization as embedded within the larger white supremacist and hegemonic project, African and African descendant people have and continue to operationalize and utilize Blackness as a means of defying physical, social, political, economic, spatial, and temporal boundaries. Black immigrants' US experiences necessitate transnational space-making, socio-political strategizing, organizing, and mobilization embedded in rhetoric and actions of solidarity and collaboration. Kadiatou Diallo (2004), in reflecting on her son Amadou's murder by the New York Police Department in 1999, explains that

> The anger expressed by people at rallies for Amadou was built on a history that I did not have and could not know, no more than they could know the specific terror of hiding in a pit, thinking that soldiers

> with guns and machetes were on their way. But the distinctions in our suffering were based more on fate than anything else. Their ancestors had been taken away to suffer the horror of slavery, and mine, through nothing but a slight accident of geography, stayed free, and lived to suffer the hardships of Africa. Our pasts and our futures now intersected through Amadou.
>
> (247)

With shared cultures, heritages, histories, and realities, it is of the utmost importance to reimagine Black immigrant mobilities and spaces and the material, media, literary, historical, cultural artifacts, and primary resources born out of them to initiate the "exploration of the new reconceptualized form of knowledge" (McKittrick 2015, 23) of Blackness in the United States. They show us, Africans and African descendants across borders, the models we've made, past and present, that reimagine the world, Black life and living. Across ethnic enclaves, Baptist churches, and HBCU campuses, in anthologies and periodicals, even in musical compositions and song lyrics, Black immigrants and Black Americans are bound to each other across centuries.

Bibliography

Adi, Hakim. 1998. *West Africans in Britain, 1900-1960: Nationalism, Pan-Africanism, and Communism*. London: Lawrence & Wishart.

Allied and Associated Powers (1914–1920). 1919. *Convention revising the general act of Berlin, February 26, 1885, and the general act and declaration of Brussels, July 2, 1890. Signed at Saint-Germain-en-Laye, September 10, 1919*. HM Stationery Office.

Bailey, Moya, and Trudy. 2018. "On misogynoir: Citation, Erasure, and Plagiarism." *Feminist Media Studies*, 18(4): 762–8.

Baldwin, Andrea N., and Mortley, Natasha K. 2020. "Caribbean Women and Reparatory Justice: Reclaiming, Rebuilding and Restoring Communities Through Migration." *International Journal of Africana Studies (IJAS)*, 21(1–2): 43–64.

Baldwin, Andrea N., Nana Afua Brantuo, and Jazmin P. Pichardo 2021. "Black Feminisms and Pedagogical Space-Making: Public Knowledge and Praxis in the Contemporary Moment." In *Handbook of Social Justice Interventions in Education*, ed. Carol A. Mullen, 1–24. New York: Springer Publishing.

Barriteau, V. Eudine. 1996. "Structural Adjustment Policies in the Caribbean: A Feminist Perspective." *NWSA Journal*, 8(1): 143–56.

———. 2001. *The Political Economy of Gender in the Twentieth-century Caribbean*. New York: Springer.

Benjamin, T. 2021. "Black Immigrant Invisibility Within Immigration Advocacy and Policy." *The Journal of the Center for Policy Analysis and Research*: 51–65. https://issuu.com/congressionalblackcaucusfoundation/docs/africa-america_2021_re-envisioning_liberation_for_?fr=sMGUzYTMyODUzNjU

Bledsoe, Adam, and Willie Jamaal Wright. 2019. "The Pluralities of Black Geographies." *Antipode* 51(2): 419–37.

Bonilla, Yarimar. 2015. *Non-Sovereign Futures: French Caribbean Politics in the Wake of Disenchantment*. Chicago: University of Chicago Press.

Brantuo, Nana Afua Y. 2017. "Targeted: Undocumented Black Immigrants Under Trump." *AAIHS*, June 26, 2017. Accessed 13 February 2022. https://www.aaihs.org/targeted-undocumented-black-immigrants-under-trump/.

Butterfield, Lyman Henry, John Catanzariti, Thomas Jefferson, and Charles T. Cullen. 1950. *The Papers of Thomas Jefferson, Volume 31: 1 February 1799 to 31 May 1800*. United Kingdom: Princeton University Press.

Byron, Margaret, and Condon Stéphanie. 2008. *Migration in Comparative Perspective: Caribbean Communities in Britain and France*. Routledge Research in Population and Migration. New York: Routledge.

Carew, Joy Gleason. 2015. "Black in the Ussr: African Diasporan Pilgrims, Expatriates and Students in Russia, from the 1920s to the First Decade of the Twenty-First Century." *African and Black Diaspora: An International Journal*, 8(2): 202–15.

Chisholm, Shirley. 1982. "US Policy and Black Refugees." *Issue: A Journal of Opinion*, 12(1/2): 22–4.

Clifton, Lucille. 1983. "Won't You Celebrate with Me." *The Book of Light*. Washington: Copper Canyon Press.

Clinton, William J. 1996. "State of the Union 1996." *Vital Speeches of the Day*, 62(9): 258.

Collins, Merle. 2020. "Louise Langdon Norton Little, Mother of Malcolm X." *Caribbean Quarterly*, 66(3): 346–69.

Convention Revising the General Act of Berlin, February 26, 1885, and the General Act and Declaration of Brussels, July 2, 1890." *The American Journal of International Law*, 15(4) (1921): 314–21. Accessed 11 May 2021.

Cooks, Carlos A. 1992. *Carlos Cooks and Black Nationalism from Garvey to Malcolm*. New York: The Majority Press.

Cresswell, Tim. 2013. *Geographic Thought: A Critical Introduction*. Critical Introductions to Geography. Chichester, West Sussex, UK: Wiley-Blackwell.

Diallo, Kadiatou, and Craig Thomas Wolff. 2004. *My Heart will Cross this Ocean: My Story, My Son, Amadou*. One World/Ballanti ne.

Dodson, Howard, Sylviane A Diouf, and Schomburg Center for Research in Black Culture. 2004. In *Motion: The African-American Migration Experience*. Washington, DC: National Geographic Society.

Doig-Acuña, Maya. 2020. "The Most Caribbean of Stories." *Southern Cultures*, 26(4): 12–23.

Dunn, Marvin. 1997. *Black Miami in the Twentieth Century*. Florida History and Culture Series. Gainesville: University Press of Florida.

Ferriss, Susan. 2020. "Hidden Hardship: Guest Farmworkers with Visas Died of COVID-19 in Obscurity While Trump Planned Wage Freezes." Center for Public Integrity, December 23, 2020. Accessed 17 January 2021. https://publicintegrity.org/inequality-poverty-opportunity/immigration/guest-farm-workers-visas-trump-wage-freezes/

Ford, Paul Leicester 1905. *The Works of Thomas Jefferson in Twelve Volumes. Federal Edition. Collected*. New York: G. P. Putnam.

Founders Online, National Archives. "From Thomas Jefferson to James Monroe, 14 July 1793." Accessed 11 May 2021. https://founders.archives.gov/documents/Jefferson/01-26-02-0445.

Halter, Marilyn. 2008. "Cape verdeans in the U.S." In *Transnational Archipelago: Perspectives on Cape Verdean Migration and Diaspora*, eds. Luís Batalha and Jørgen Carling, 35–46. Amsterdam University Press. http://www.jstor.org/stable/j.ctt46msd4.6.

Harris, LaShawn. 2008. "Playing the Numbers: Madame Stephanie St. Clair and African American Policy Culture in Harlem." *Black Women, Gender & Families*, 2(2): 53–76.

Harris, Wilson. 1965. *Tradition and the West Indian Novel: A Lecture Delivered to the London West Indian Students' Union on Friday, 15th May, 1964*. London: London West Indian Students' Union.

Hilal, M. 2022. *Innocent Until Proven Muslim: Islamophobia, the War on Terror, and the Muslim Experience Since 9/11*. Minneapolis, MN: Broadleaf Books.

Hudson, Peter James. 2017. *Bankers and Empire: How Wall Street Colonized the Caribbean*. Illinois: University of Chicago Press.

Johnson, Buzz. 1985. *"I Think of My Mother": Notes on the Life and Times of Claudia Jones*. London: Karia Press.

Journal of the Annual Session of the National Baptist Convention. United States: National Baptist Publication Board, 1899. https://www.google.com/books/edition/Journal_of_the_Annual_Session_of_the_Nat/CWEyBOyf9RkC?hl=en&gbpv=0

Kennedy, Paul M. 1989. *The Rise and Fall of the Great Powers: Economic Change and Military Conflict from 1500 to 2000*. New York: Vintage Books.

Kincaid, Jamaica. 1988. *A Small Place*. 1st ed. New York: Farrar, Straus, Giroux.

King, Tiffany Lethabo. 2019. *The Black shoals*. Durham: Duke University Press.

Lima-Neves, Terza Alice Silva. 2015. "D'NOS MANERA-Gender, Collective Identity and Leadership in the Cape Verdean Community in the United States." *Journal of Cape Verdean Studies*, 1(1): 57–82.

Lundy, Garvey F. 2006. "Early Saint Domingan Migration to America and the Attraction of Philadelphia." *Journal of Haitian Studies*, 12(1): 76–94.

Marable, Manning. 2000. *How Capitalism Underdeveloped Black America: Problems in Race, Political Economy, and Society*. South End Press Classics, V. 4. Cambridge, MA: South End Press.

Martin, James W. 2018. *Banana Cowboys: The United Fruit Company and the Culture of Corporate Colonialism*. Albuquerque: University of New Mexico Press.

McKittrick, Katherine, and Clyde Adrian Woods. 2007. *Black Geographies and the Politics of Place*. Toronto: Between the Lines.

McKittrick, Katherine. 2021. *Dear Science and Other Stories*. Durham: Duke University Press.

———. 2015. *Sylvia Wynter: On Being Human as Praxis*. Durham: Duke University Press. https://read.dukeupress.edu/books/book/199/Sylvia-WynterOn-Being-Human-as-Praxis

———. 2006. *Demonic Grounds: Black Women and the Cartographies of Struggle*. Minneapolis: University of Minnesota Press.

McLean, Shay-Akil. 2021. "Patriarchy & Gender." Decolonize ALL The Things. Accessed 11 January 2022. https://decolonizeallthethings.com/learning-tools/patriarchy-gender-lesson-plan/.

Morrow, Diane Batts. 2002. *Persons of Color and Religious at the Same Time: The Oblate Sisters of Providence, 1828–1860*. Chapel Hill: University of North Carolina Press.

Newman, Richard. 2014. *This Far By Faith: Readings in African-American Women's Religious Biography*. New York: Taylor & Francis.

Olufemi, Lola. 2021. *Feminism, Interrupted: Disrupting Power. Outspoken*. London: Pluto Press.

Palmer, Breanne J. 2017. "The Crossroads: Being Black, Immigrant, and Undocumented in the Era of# BlackLivesMatter." *The Georgetown Journal of Law and Modern Critical Race Perspectives* 9: 99.

Patriarchy & Gender. 2015. “Decolonize ALL The Things.” Accessed 21 March 2021. https://decolonizeallthethings.com/learning-tools/patriarchy-gender-lesson-plan/

Perry, Imani. 2004. *Prophets of the Hood: Politics and Poetics in Hip Hop*. Durham: Duke University Press.

Plummer, Nellie Arnold. 1997. *Out of the Depths, or, the Triumph of the Cross. African-American Women Writers, 1910-1940*. New York: G.K. Hall.

de Reid, Ira A. 1938. “Negro Immigration to the United States.” *Social Forces*, 46(3): 411–7.

Reid-Maroney, Nina. 2004. “African Canadian Women and the New World Diaspora, c. 1865.” *Canadian Woman Studies* 23(2): 92–6.

Riggs, John Beverley. 1946. “Certain Early Maryland Landowners in the Vicinity of Washington.” *Records of the Columbia Historical Society, Washington, DC*. 48: 249–63.

Robinson, Cedric J. 2005. *Black Marxism: The Making of the Black Radical Tradition*. Chapel Hill: Univ of North Carolina Press.

Rodney, Walter, A. M. Babu, and Vincent Harding. 1981. *How Europe Underdeveloped Africa*. Rev. pbk. ed. Washington, DC: Howard University Press..

Roosevelt, Theodore. 1904. “State of the Union Address” (speech, December 6, 1904), Theodore Roosevelt Center. Accessed 15 June 2022. https://www.theodorerooseveltcenter.org/Learn-About-TR/TR-Encyclopedia/Foreign-Affairs/Roosevelt-Corollary

SAPRIN (Organization), World Bank, and Citizens' Assessment of Structural Adjustment (Organization). 2002. *The Policy Roots of Economic Crisis and Poverty: A Multi-Country Participatory Assessment of Structural Adjustment: Executive Summary*. Washington, DC: SAPRIN.

Songwriters Hall of Fame. “Andy Razaf.” Andy Razaf | Songwriters Hall of Fame. Accessed 11 May 2021. https://www.songhall.org/profile/Andy_Razaf.

Span, Christopher M., and Brenda N. Sanya. 2019. “Education and the African diaspora.” In *The Oxford Handbook of the history of education*, eds. John L. Rury and Eileen H. Tamura, 399–412. Oxford: Oxford University Press.

Spelman Messenger. 1915. “Spelman in Africa, Africa in Spelman.” 32(2). Accessed 11 March 2021. http://hdl.handle.net/20.500.12322/sc.001.messenger:1915.07

Strickland, Scott M., Virginia R. Busby, and Julia A. King. 2015. “Indigenous Cultural Landscapes Study for the Nanjemoy and Mattawoman Creek Watersheds.” *Report to National Park Service Chesapeake Bay Office, Annapolis, MD, from St. Mary's College of Maryland, St. Mary's City*.

Structural Adjustment Participatory Review International Network. 2004. *Structural Adjustment: The SAPRI report; the policy roots of economic crisis, poverty, and inequality; a report on a Joint Participatory Investigation by Civil Society and the World Bank of the impact of structural adjustment policies*. Zed Books.

Summers, Brandi Thompson. 2019. *Black in Place: The Spatial Aesthetics of Race in a Post-chocolate City*. Chapel Hill: UNC Press Books.

Tamale, Sylvia. 2020. *Decolonization and Afro-Feminism*. Ottawa, ON: Daraja Press.

The Institute Monthly. 1912. “Are the African Youths in Training Worth the Cost?” 4(7). Accessed 11 May 2021. http://library.wvstateu.edu/archives/college_publications/Institute-Monthly/1912-01.pdf

United States Government. Central Intelligence Agency. 1980. Cuba: Educating Future Third World Leaders. Accessed 11 May 2021. https://www.cia.gov/readingroom/docs/CIA-RDP91-00965R000200020027-7.pdf

United States Government. Central Intelligence Agency. 1970. Black Radicalism in the Caribbean - Another Look. Accessed 11 May 2021. https://www.cia.gov/readingroom/document/cia-rdp85t00875r001100090030-4

———. 1964. Cuban Training and Support for African Nationalists. Accessed 11 May 2021. https://www.cia.gov/readingroom/document/cia-rdp79-00927a004300100001-5

United States. 1886. Department of State, and Berlin West Africa Conference (1884–1885: Berlin, Germany). Message from the President of the United States, Transmitting a Report of the Secretary of State Relative to Affairs of the Independent State of the Congo. [United States] 49th Congress, 1st Session. Senate. Ex. Doc, No. 196. Washington: Government Printing Office.

Vialet, Joyce C. 1978. *The West Indies (BWI) Temporary Alien Labor Program, 1943–1977: A Study*. US Government Printing Office.

Vine, David. 2020. *The United States of War: A Global History of America's Endless Conflicts, from Columbus to the Islamic State*. Berkeley, CA: University of California Press.

Wynter, Sylvia. 2007. "Human Being as Noun? Or being human as praxis? Towards the autopoetic turn/overturn: A manifesto." *Unpublished essay*.

8 The Women. They Were Plotting Too

Declaring Our Independence in the Spirit of Sankofa

Barby Asante

The Opening Scene

I speak in the tongues of my mother and my grandmother and my grandmothers before her
Although I do not know the words
My very being is the evidence of those languages
It is in these words I will speak
It is in these words that I will create
It is in these words that I will question

I question this place, a Diaspora place
as an ideal or narrative
Of my relationship to my internal migratory ROOTS
As the distance between becomes more apparent as we get to 2nd, 3rd & 4th generations removed from "M/otherlands"

How do I map myself in this place?

my family, the people that left
and arrived
and continue to arrive,
even though they make borders,
make checkpoints
and build walls

The process of making
Is enacting my presence
The hard navigation of presences,
absences, visibility, invisibility[1]

DOI: 10.4324/9781003143550-12

Introduction

The scene that opens chapter is from one of the first *Declaration of Independence* performances. This iteration was born from the grief and anger at the death of Khadija Saye and her mother Mary Mende in the Grenfell Tower Fire in 2017. At the time, I had recently met Khadija, as we were both showing work in the Diaspora Pavilion in Venice. Our work shared a space in the study of the Palazzo Pisana Santa Marina that housed the Diaspora Pavilion during the 57th Venice Biennale. *Declaration of Independence* is dedicated to Khadija Saye, Mary Mende, my grandmother Dina Kwansema Edwin Baiden (Aba) and *"for the women who contribute to me, inspire me and keep me alive. My mothers, lovers, sisters, cousins, daughters, comrades, co-conspirators."*[2]

This chapter considers the creation of *Declaration of Independence*—how it is grounded in an ethics of care rooted in the West African Akan principle of *Sankofa* and the particularities of a transcultural Black feminist thinking and practice that takes its position from my being of Ghanaian parentage born in the United Kingdom. This positionality has historic and geographic arrangements that allow for a convergence of many Black feminist propositions, which come from being in a place of one of the most prevalent imperial powers in the world, with a deeply persistent nostalgia for this imperial past and assumed greatness. This place, the United Kingdom is in close proximity to other imperial rivals some of whom have come closer to facing their imperial defeat even as their reliance on facets of this supremacy persist, subtly and not so subtly. This island place, that has experienced many invasions, invaded 90% of the world and colonized 25% of it. *We*[3] arrived here, by invitation. Then out of economic necessity. And now we have stayed, given birth and many others have arrived, and continue to arrive, sometimes invited, sometimes out of economic necessity, sometimes for refuge. The perception is that *we* are some sort of *invasion* and not of our belonging and intimate connection.[4]

Saidiya Hartman's piece for the Feminist Art Coalition, *The Plot of Her Undoing* (2019), reveals a plan with many beginnings and continuations that impact the lives of Black women. Her text weaves a story of violence and domination steeped in the enduring legacies of slavery, colonialism, and white supremacy. Reading this piece, I heard resonances with *Declaration of Independence*. When Hartman shifts the focus of the piece from revealing the *plot* that *un-does* us to revealing a trajectory for the *plots undoing*, she is suggesting that it "proceeds by stealth" (5). It is an *undoing* that "is almost never recognized as anything at all and certainly never as significant" (5) and "is not for your entertainment, even if it is for your benefit" (5) implying that the work of undoing is for those who need these conditions to be *undone*. This work is not entertainment in so much as although it may seem increasingly visible, like in the ways in which videos of black people being killed by law enforcement have been captured and distributed, or as in *Declaration of Independence*, I invite Black and womxn of color to share their stories and experiences of living in a world where the legacies of coloniality and enslavement continue to color the lives they live. Although this work has been created and performed in public art galleries and international artistic platforms, this work is for us and by us. It is an artwork about the coming together of Black and womxn of color[5] to share ourselves and our struggles so that we might "pledge our love" and share our strategies for the *undoing*. *Declaration of Independence* strives to *undo* the prevailing injustices of slavery, colonization, and white supremacy veiled unsuccessfully within the discourse around the development of more diverse art spaces, fueled by an increased awareness of injustice and institutional inequity driven by the clarion call for decolonization and abolition.[6] Within the arts organizations and other artistic platforms I have been invited to present the work in, *Declaration of Independence*, seen as a project that visibly represents the kind of *diversity* and anti-racist position that such spaces wish to present as an antidote to their implication and complicity in these legacies. But *Declaration of Independence*

> *…is our process,*
> *it is not about changing others,*
> *but rather an invitation for us to change our relationship to ourselves, to our communities.*
> *and to the wider world.*[7]

Declaration of Independence is a refusal to be complicit. It is an *undoing* of the plot that acknowledges the many *undoings* that have brought me and more than 70 Black and womxn of color, who have taken part in the project since 2017, to where we are today and into those spaces to disrupt. We are the "*mothers, lovers, sisters, cousins, daughters, comrades, co-conspirators.*"[8] Our presence in these spaces is as much about the visibility that these "diversity" initiatives offer as it is about our refusal of the *tokenistic* gesture toward a historical reconciliation. In this performance work, we come together in a powerful and emotive presentation of ourselves, bringing with us our anger, our grief, and our celebration in a ritual circle that is as much a resource for us as it is an artwork.

Declaration of Independence

What is Declaration of Independence? It is an iterative, ensemble performance first performed at the opening of the Diaspora Pavilion in Venice in 2017. Although it was not named *Declaration of Independence* at that point, the shape of the performance was developed as a live accompaniment to *Intimacy and Distance*, an installation intervention created in dialogue with a chosen group of Black and womxn of color from my family and kinship circle. *Intimacy and Distance* considers the intimacies and distances between us and the places we are geographically connected to. It was an exploration of our historical relationships to the displacements of diasporas and how these are intimately connected to Venice as a floating city known for its maritime power and its centrality to the historic development of imperialism, global trade, and thus the project of modernity.[9] I wanted to scrutinize the position of the Diaspora Pavilion at the Venice Biennale, a platform that exists to celebrate the artistic virtuosity of an individual artist representing a nation-state. I also wanted to bring to the domestic space of the Palazzo Pisana Santa Marina, appropriated for use as an unofficial pavilion,[10] something of what Jacqui Alexander describes as "radical [and intersectional] feminist politics" (2006, 257).

The invitation to my friends and family was to contribute their thoughts, words, and voices to the work through a recorded Skype conversation based on *Poem for Aba*, a poem I wrote for my grandmother at the time of her death, and on Ama Ata Aidoo's poem *As Always, A Painful Declaration of Independence* (1992). I will expand on the significance of these pieces of writing in more detail later in this chapter. I also asked that each contributor give me instructions for rituals that I would perform and film myself doing on their behalf in and around the Palazzo. My intention was to leave traces of these rituals with some carefully chosen objects that had small screens embedded within them playing videos of those rituals. I placed small discreet speakers in corners, under tables, in the courtyard, in a bedroom, bathroom, and on the bookshelves of the study. These speakers wove womxn's voices in echoes, whispers, and shouts throughout the space in a quiet but very present fragmented composition around the Palazzo. The conversations and responses to the poems that scored this composition were testimonies, declarations of anger, joy, hopes, dreams, and most of all, our precarious presence in the world. It seemed a logical next step to have these voices presence themselves at the opening of the performance. This performance was an opportunity to declare ourselves, in Venice, as an alive and living embodiment of the histories and present of an ongoing coloniality that has roots/routes in the development of Venice as a center of trade, map-making, and maritime exploration.

Key to the *Declaration of Independence undoing of the plot* is the possibility of imagining and enacting an *otherwise* way of being. I borrow the definition of otherwise from an event at the Utrecht-based arts, social action and research organization, BAK, basis voor actuele kunst, where I presented a performative lecture version of *Declaration of Independence* as part of their Propositions #10: Instituting Otherwise symposium in 2019. This symposium was convened at the time of *Trainings for the Not Yet*, a program of interventions and working groups curated by Dutch artist and curator Jeanne Van Heeswijk in collaboration with the organization. The description of *otherwise* proposed was "the actualization of the collective alternative imaginings of the 'not-yet': of 'another world' and 'another future.'"[11] I will add to this definition the possibility of using other ways of knowing and understanding, namely the Akan Adinkra principle of *Sankofa*. It is symbolized by a pictograph of a bird with its feet facing forward and its long neck facing backward toward an egg that is resting on its back. *Sankofa*'s loose translation from the Akan languages of Twi and Fanté is "retrieve" or "go back and get" (Asante and Mazama 2009, 7). As well as being significant to indigenous Akan cultures, *Sankofa* has

Figure 8.1 Declaration of Independence Performance, Diaspora Pavilion, Venice, May 2017. Photo credit Francesco Allegretto.

Figure 8.2 Intimacy and Distance Installation Extracts, Diaspora Pavilion, Venice, May 2017. Photo credit Francesco Allegretto.

significance across the African Diaspora, especially in the United Kingdom and the United States through African nationalist and Pan-African consciousness raising and organizing. *Sankofa* has been employed as something that represents the need to reflect on the past in order to move forward, particularly in relation to the legacy of the transatlantic slave trade, coloniality, and the contemporary situation of "underdevelopment" (Rodney 2018) that persist in Africa as a result of colonization. Being of Akan heritage, I grew up hearing the words *San-ko-fa* used daily to remind my sister and I to go back and get whatever we forgetful children had left in our bedroom, in the supermarket or in the cloakroom at school. We heard it even if we forgot each other, which as sisters in pursuit of other friendships, we sometimes did. I am proposing a re-reading of *Sankofa* from a Black feminist perspective in the context of a diasporic geographic situation that not only returns to "go back and get" but also considers the importance of a remembering that is collective and embodied. This reading not only wants to return to the past to contest or redress the "facts" of history but also thinks about how we care for what we remember and how we use memory as an everyday practice of care for that which is to come.

My Black feminist position comes from the many intersecting linkages that span the African Diaspora and land in this place, in my mind, my spirit, and my body on this island. Black feminisms from North America provide me with modes of analysis to support emerging local understandings of Black womxn's experiences and their social, political, and creative work toward liberatory politics, practices, and ways of knowing. I also draw on my personal experience growing up in a matrilineal Akan culture, where women are held in high regard and hold power, but where patriarchal tradition, informed by Akan traditional religion (Asante and Mazama 2009, 24) and colonialism (Maja-Pearce 1990, 17), prevails; African women and African Feminist movements continue to address these traditions. To create a central triad of significant influences, I also call upon the Black British feminist organizing by Brixton Black Women's Group (BWG) and the Organization of Women of African and Asian Descent (OWAAD), which I was not part of, but whose legacies resonate deeply in contemporary Black British and womxn of color theorizing, activism, and creating. It is important to add that Black feminist activists in Britain during the 1980s used Black as a political term that brought together Black African and Caribbean, South Asian, and other minority groups organizing in anti-racist, anti-imperialist, and anti-capitalist struggles, locally and globally. This was a significant and important organizing strategy and as contested then as it is now, especially with the increasing awareness of the impact of anti-blackness and its social, psychological, and political impact on people of African descent around the world. While I recognize the need to contest the dynamics of race that trouble organizing in coalition, I organize *Declaration of Independence* in a coalition of Black and womxn of color. I do this in honor of the sisters who came before us, their relentless efforts to imagine another future in which I am present living, working, and studying, and as a continuation of this imagining as united fronts. In order to continue, we might need to learn how to be together again with all that is discomforting to us. These are my central feminist linages. They intersect and contradict each other, but essentially, they represent a Black feminism that challenges the prevailing order of things. These references provide the guidebooks, instructions, maps, and manifestos that I draw on to create *Declaration of Independence*.

Dramaturgy

Declaration of Independence is an act of speech, testimony, and proposition that draws on the American Declaration of Independence (1776), the revolutionary declarations of the Zapatistas (1993–2005), the Combahee River Collective Statement (1977), and the Charter of Principles for African Feminists (2006). In a recent podcast series[12] developed with womxn who took part in the project in Bergen Kunsthall in Norway, I described the project as a resource of study, kinship, and connection, sharing information, sharing space, and most of all sharing ourselves. I use an artistic strategy of *study* drawing on the role that study groups played in the consciousness-raising and movement-building in groups such as the Black Panthers and the Brixton Black Women's Group (Lewis et al. 2017). I also engage Stefano Harney and Fred Moten's (2013) extension of this definition of *study* into a political motivation, which has both intellectual and non-intellectual motivations as part of Black social life. This theorizes *study* as "what you do with other people," which could be "talking and walking around with other people, working, dancing, suffering, some irreducible convergence of all three, held under the name of speculative practice" (Class War University 2012). *Declaration of Independence* is steeped in this *speculative* approach that draws as much on everyday togetherness as it does on the creation of an event or occasion. The *Sankofa* within it adds a *speculative* possibility grounded in remembering our everyday interactions and imaginings of the future as well as our reflection on what has come before.

The production of this work as *study* is an important part of the dramaturgy of *Declaration of Independence*.[13] For this chapter, I draw on quotes from performance texts and surrounding materials of *study*. This is an intentional choice to invite you (the reader/the audience) into the performative process of the *Declaration of Independence*, and the ways in which the performance opens with a call from me that invites a response from those who become part of the project. This call is also the call of many who came before us, and our presence in life is the ever-spiraling response. These quotes and references refer to the ways in which *Declaration of Independence* draws on Ama Ata Aidoo's poem *As Always, A Painful Declaration of Independence* (1992) as a core text in our collective process of *undoing*. I will reflect on the creation of this work, which happened in art contexts until late 2020,[14] using spaces and resources to make a container "for us."[15] I will then move into how *Declaration of Independence* is a performance but also a ritual to create community and resources. This approach will look at the way the performance is formed as *study* and the ways in which we come together and form a circle

Figure 8.3 Declaration of Independence Workshop, Bergen Kunsthall, Bergen, December 2019. Photo credit Nayara Leite-Day.

that facilitates the collective creation of the work. In particular, the way that each person contributes to the whole through collective and individual writing that builds on and advances the knowledges that are available for our *study*. Lastly, I will explore how *Sankofa* occurs in the work. I will think about the ways in which our histories and geographies intersect and weave in and through the work in its different iterations, and how these are met and witnessed in what can be described as a rememory[16] that spirals across time space and the different iterations of the performance. It is my hope that within these spirals, there is space for some sort of transformation to occur. I am hoping that this chapter will shed light on the way I am conceptualizing the work and what I am bringing with me to create the container for the performances. I can only attest to my own experience of the work within this text, but with so many having witnessed, participated in, and contributed to the work, I'm sure there are many understandings and experiences of its transformative potential yet to be shared.

On-doing Undoing

Ama Ata Aidoo's poem *As Always, A Painful Declaration of Independence* (1992) declares independence, not only from colonizers but also from the racial and gender-based repressions that persist as a legacy of coloniality. Mapping a trajectory of *on-doing*[17] that resonates with the *undoing* expressed in Saidiya Hartman's essay, the poem begins in the fifteenth century with "a philosopher at his desk, pen in hand, as he sorts the world into categories of genus and species" (Hartman 2019, 1). Ama Ata Aidoo knows the beginning of this plot intimately. She knows that she has been taken in. She knows that she once trusted this relationship. That this man who landed in Saltpond, Cape Coast, and Elmina[18] in his ships, filled with his guns and pacotille, who built forts and castles for his own profit and protection; she knows how this beginning was a series of betrayals that led to the extractive trading of gold and people, which began in what we now know as Ghana in the 1480s. First, the man was Portuguese, followed closely by a Dutchman, a Swede, some Danish, some Germans, and then the British ones, who in 1847 established control over the territory in which she lived and called it The Gold Coast. She thought that being in a relationship with him might be beneficial to her development, but the trade in gold and people made the European men wealthy and left Africans bereft and wanting. However, the European possession of her land has not been so easily surrendered. There has been war and anti-colonial struggle and attempts at *undoing the plot* of imperialism and colonization. The independence of Ghana that was fought for, negotiated, and granted by the British in 1957 did not yield the self-determination and liberation that was hoped for.

As Ama contemplates how to *undo* this *plot* she writes:

Friend
height
was not missed much
except when

I needed to reach the
upper shelves of my existence
from where I would want
-occasionally –
to bring down
my forgotten
hopes
aspirations,
plans and projects
to be dusted…[19]

Directing her words to the colonizer/ the philosopher/ this supposed benefactor, a lover, a brother and a maybe comrade, the piece alludes to a disappointing love affair—an affair that limited her ability to thrive rather than recognizing her humanity or encouraging reciprocity. It is an affair so steeped in betrayal that their connection and intimacy is fervently denied by him. This realization precedes the *on-doing of the plot* for Ama. It is the point at which "…everything has been taken. When life approaches extinction, when no one will be spared, when nothing is all that is left, when she is all that is left" (Hartman 2019, 5). Ama chooses herself because this is all she has and severs the painful and tempestuous relationship that has taken away all possibility of a self-determined life:

Oh My Brother,
the decision to
sever and separate
was not based on the knowledge of me
as a forever dreamer
unable and incapable of handling
the clear world of
take and
take and
take[20]

Ama does not know where this separation will take her. She just knows that she can no longer show up.

So my dear, my love,
I'm gone.
I'm through[21]

As Always, A Painful Declaration of Independence has been shared with every womxn who has contributed to this project since its genesis in 2017. It is the core text of *Declaration of Independence*, and acknowledges that I, that we, can no longer handle this "clear world of take and take and take" (Ata Aidoo 1992, 9), that we are all that we have left, and that we need something else. *Declaration of Independence* makes space for Black and womxn of color to presence themselves, be witnessed, seen, and to tell their stories. It is an attempt to hold space for discontent and disappointment, for the promises not fulfilled, for the promises of a better life, of self-governance and self-determination, education for our children, jobs for us, healthcare, and streets paved with gold.

As an iterative performance I imagine *Declaration of Independence* as a transient generative space, which is both a forum and a circle. I draw on the notion of forum to recognize the politics and FORM-ality of our coming together as well as a space to reflect on the kinds of state agreements, treaties, and policies that affect our lives past, present, and future. As an iterative performance, I imagine *Declaration of Independence* as a transient generative space, which is both a forum and a circle.

In 2019 *Declaration of Independence* was presented at the BALTIC Centre for Contemporary Art in Gateshead in the United Kingdom. This invitation presented the possibility of not just a one-off performance but also to think about *Declaration of Independence* in the context of an exhibition space. I wanted to create an installation that alluded to our presence in the space and invited the viewer into the center of that experience, evoking as much as possible the kind of conditions generated in a live performance. I asked myself, how could I present the concept of our circle within the static space of a gallery exhibition? How could I evoke the essence of our performative declaration for a gallery audience? How could this space also be a space of reflection and resource for others? Working with Black Dutch architect Afaina De Jong and producer Jess Harrington, we created a space that would present *Declaration of Independence* as a gallery exhibit which presented our performance as a looping video work while also being a performance space for us to present a live *Declaration* performance for the opening of the exhibition. The space was also created with the view that other groups of people could arrange with the gallery team to use the space to explore their own declarative processes and propositions.

The design drew on references from the texts of previous performances along with archival materials depicting spaces created for conferences and forums, such as The Bandung Conference, the first large-scale meeting of African and Asian states in 1955, and the subsequent first Non-Aligned Movement Summit in Belgrade in 1961 convened to promote African and Asian alliances for economic and cultural cooperation and to oppose colonialism and neocolonialism. Because our space was to center the feminist experience of womxn and Black and womxn of color, we also referenced the Conference for Women in Africa and the Diaspora organized by the Ghana women's movement, Becoming Visible, the first Black Lesbian conference held in the women's building in San Francisco in 1980, and other such women's conferences. We looked at the spaces, logos, and other graphic materials not just around conferences, but also those of protest and movement building, which were anti-colonial and womxn centered. The resulting space was configured in the round and imbued with a spirit of these FORM-alities. We designed a project logo, which was also not exactly a logo in the conventional sense. Rather it was an arrangement of circles and triangles in bright and pastel colors that intersected in different configurations presenting an illustrative representation of the ways in which our circle changed, transformed and evolved. Two circular benches painted in blue and yellow and described by Afaina as a circle of peers and a circle of witnesses filled the gallery space. The brightness and informality of this furniture provided adequate seating for the performers and the audience when we presented the live declaration while also serving as an invitation for gallery audiences to sit and watch the "virtual" *Declaration of Independence* that unfolded on three large screens, organized as if a live broadcast was taking place in the space. We had no idea that a virtual *Declaration of Independence* would become a reality in 2020.

This Is Our Circle

Declaration of Independence does not merely function as a spectral opportunity for the mostly white audiences of the arts organizations, in which the work has been presented, to see Black and womxn of color speak their truths. Rather, it is deeply grounded in the idea of forum, community, and circle as a virtual space that is created to hold us together. The circle is not open, and the forum is not open for comments, although during the performance there is an

Figure 8.4 Declaration of Independence Performance, BALTIC Centre for Contemporary Arts, Gateshead, March 2018. Photo credit Colin Davis.

unspoken invitation for other Black and womxn of color to join. Unspoken in that is not explicitly articulated, but the invitation is in the words we speak. This circle is not open in order to protect the space created for us to tell our stories. It is closed so that the audience become witness to our circle process, so they hear us without invalidating or questioning our declaration. The audience is held in our gaze as we speak our words. We are in the position of power. Our circle has been strengthened and fortified by our togetherness and its connection to the work of other womxn who have come before us and previous iterations of the project. The words we write and speak are our living collective narratives, spoken into reality. These words are the words of our ancestors reborn through the documents that are our bodies:

I speak in the tongues of my mother and my grandmother and my grandmothers before her
Although I do not know the words
My very being is the evidence of those languages
It is in these words I will speak
It is in these words that I will create
It is in these words that I will question[22]

At the end of each performance, the circle is opened in the way we move toward the audience to invite them to dance with us and join us to celebrate our declaration.

Many of the people who have been part of this project have never taken part in a performance. Some have some experience of public speaking, but the level of openness and vulnerability that comes from making such a declaration in public is not something that comes as easy as delivering a paper, presentation, or motivational speech. As *Declaration of Independence* is formed within layers of difficult contexts, particular attention is paid to the ways in which we come together in order to create the circle. I draw on my experience of circle processes from sister and healing circles to the kinds held in activist spaces to facilitate difficult and transformative moments. African communities have used circle processes for centuries, for disputes, peace-making, initiation, rituals, and more. Forming the circle for *Declaration of Independence* has been an important process to enable us to be of service to whatever comes up within the space—to allow someone to show up as their whole self, let their guard down, and have someone else hold and witness them. To facilitate this, I also become the container for their expression. I am the facilitator, the dramaturge, the master of ceremonies, and the host of the space, making sure that the experience is a comforting and nurturing one.

To say then that *Declaration of Independence* is formed in a "workshop" does not do justice to what we all bring to be together and to bring a performance into being, hence the idea of being in a process of *study*. For example, for the 2018 Library of Performing Rights performance at the Live Art Development Agency, London, I meticulously planned how the performance would unfold in a process of togetherness. This included where we would be together and how we would nourish this togetherness with food and other activities before we began to devise the performance. We did somatic body work, clearing rituals, collective reading, a sound bath, dancing together, laughing together, and crying together. All of this feeds into the collective writing process of developing a performance script. Although many parts are written and performed by an individual, they are part of the whole and a reflection of the process of coming together. The importance of this approach became explicitly clear when writer and therapist Foluke Taylor, a contributor to *Declaration of Independence* and long-time friend, and I opened the space for the development of a performance hosted by Bergen Kunsthall in Norway. When the group came together and began to witness each other, there was a palatable expansion of the air in the gallery space as the womxn in the room relaxed, opened their hearts and breathed gently together. They shared their stories, recognizing themselves in each other. For many, this was the first time that they could speak freely about their life experiences as migrants to Norway.

Declaration of Independence has manifested many projects, collaborations, partnerships, love relationships, and other possibilities. One such powerful experience came when one of the younger womxn was left in Mogadishu by her mother. She was left in a place masquerading as a school but was in fact a facility where "wayward" women and children were left in order to be reformed in the Islamic way. The togetherness, care, and at stake-ness that had been generated through this project was of great help and support to her concerned sister, as we were able to assist her by generating money and contacts that helped facilitate her safe return.

In a conversation following a performance in London, Dr. Karen Salt[23] described to me her experience of seeing the performance as a process of spiraling. As an individual performer spoke their words, these words echoed through time and space, circling the performance space, and creating a spiral of energy. Drawing on the Spiralism as theorized by Haitian writers René Philoctète, Jean-Claude Fignolé, and Frankétienne, Karen describes witnessing *Declaration of Independence* as something that is ritualized in a spiritual sense while also being a strategy of survival. Relating this to how Frankétienne describes Spiralism in the opening of his novel *Ready to Burst* as: "[a] definition of life at the level of relations (colors, odours, sounds, signs, words) and historical connections (positionings in space and time)" (2014, 7). Sprialism connects art, literature, and life in a way that "novelistic description, poetic breath, theatrical effect, narratives, stories, autobiographical sketches, and fiction all coexists harmoniously" (2014, 7). Dr. Salt saw this living, spiraling energy in the performance as a form of knowledge, which is both generated through the performance and remembered through the performative process. It is an interruption of a performance space where we are simultaneously visible/vulnerable and held. *Declaration of Independence* then becomes a place to remember ourselves and a refusal of the notion of independence. It's a declaration of the interconnectedness of our stories across time, and our need for each other to support our *undoing of the plot* in order to instigate the actions to bring about the transformation we want to see in our worlds.

The Bird Is Singing Us an Invitation to Meet Our Future

She took her songs with her
Along with her stories
Her recipes
Her jokes
Her grooming techniques
Knowledge of the before, before.
Before the men started plotting the downfall of the British.
They didn't know that the women
The Women
They were plotting too![24]

The title of this section references a workshop I presented at the ICA in London in August 2019, as part of the Black Quantum Futurism: Temporal Deprogramming Residency. The workshop explored the ways in which I connect fragments of memory within my work, refusing the authority of the archival document in favor of other ways of exploring and understanding memory. Prior to *Declaration of Independence*, I had made many other works that explore memory and remembering, within the context of a country that relegated the remembering of our histories to the month of October- Black History Month. These works were expressions of my experience of working on archiving projects with groups of elders and young people. I had to come to terms with the fact that these gestural invitations by arts organizations and archives to bring groups into their spaces to explore black memory often did not wish to hold Black memories within their collections. What seemed important in these projects is that we were visibly present within the program of these organizations, so they could show the world (and their funders) that they were doing their diversity work. This frustrated me. Our significant events, our everyday interior lives and our relationships to the continuing legacies of enslavement and colonialism seemed irrelevant to them. The spiraling energy that Dr. Salt experienced in her witnessing of a *Declaration of Independence* is something of how I wanted to attend to and create an active and attentive space for remembering, conjured through the re-membering that is inherent in the ways in which the performance repeats and recalls through the words, gestures and the bodies of the performers.

In *Beloved* (1987), Toni Morrison introduces us to the idea of rememory through the recollections of Sethe, the main character haunted by the ghost of her dead daughter. She writes, "What I remember is a picture floating around out there outside my head. I mean, even if I don't think it, even if I die, the picture of what I did, or knew, or saw, is still out there. Right in the place where it happened" (36). The concept of the memory being still out there, "A thought picture" (Morrison 1987, 36) that never goes away. Even death or destruction cannot erase the memory. It can be a memory that belongs to someone else, yet "…if you go there—you who never was there—if you go there and stand in the place where it was, it will happen again; it will be there for you, waiting for you" (Morrison 1987, 36). Rememory is like the spiral. It creates wholes from "mere details and secondary materials" (Frankétienne 2014, 7) as a way to bring to light things often "too terrible to relate" (Morrison 2019: 238). In her essay *Rememory as a Strategy of Subversive Representation: A Feminist Reading of Morrison's Beloved* (2013), Madhumita Purkayastha describes Morrison's reinvention of memory as "rememory" as a "narrative strategy" that puts forward a "counter-hegemonic storytelling" where multiple perspectives converge. She describes this as being within a Black feminist literary tradition that stresses the need for "alternative reconstructions of the past and narrative strategies of subversive representation in articulating the anguish and trauma of slavery, repressed memories and tales impossible to tell" (2). Rememory thus becomes the "other side" of the story. A strategy by which that which has not been told or is unspeakable can make itself present. This spiraling rememory, conceptualized in different places in the African Diaspora, has its roots in African cultural retentions that think about memory as a pathway to attend to that which history cannot and does not contain in its *plot*. These remembering strategies are both literary and ritual ways to reconcile what has been lost and to understand what needed to happen for us to be here.

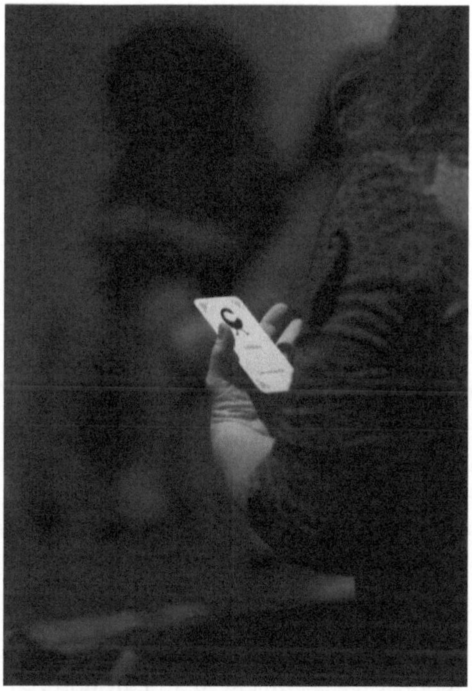

Figure 8.5 That Bird Is Singing Us an Invitation to Meet Our Future, Black Quantum Futurism, Temporal Programming, ICA, London, August 2019. Photo credit Myah Jeffers.

I opened this section with an extract from Poem for Aba, a piece I wrote for my grandmother Dina Kwansema Edwin Baiden (Aba), on the event of her passing in 2016 after 102 years of living. Poem for Aba asks how can I "go back and get" her library of knowledges? How can I reconcile the intimacy and distance that lives between our lives? I had barely known her, but a "thought picture" (Morrison 1987, 36) of her kind face is etched in my mind, and an echo of her voice speaking affectionate words in Fanté whispers to me as I walk in this world. At the time of her funeral rites, I learned from the sharing of family memories during the church services, ceremonies, and rituals, that this woman, my grandmother, prayed for us all every night. In her prayer, she called the names of each of her children, those living and those that have passed. She called the names of every grandchild and every great-grandchild. This action of calling our names was her declaration of her love for us, whether she had met us or seen us in a faded photograph or pixelated digital image. In this gesture, she declared that she loved us all, dreamed for us all, and remembered us all.

I call my grandmother along with Khadija Saye and her mother in the opening of this essay. Their memory along with the memories of many other people, events, occasions, and other things live within the words uttered, silences, tears, laughter, and movements that make up *Declaration of Independence*. I like to think that inherent to *Declaration of Independence* is the spirit of *Sankofa* imparting a way of understanding how we can use memory to imagine and nurture the future.

Sankofa is a principle of Akan cosmology belonging to the Adinkra system of "communicators" (Temple 2009, 130). Made up of pictorial symbols illustrating proverbs, which express the cultural, spiritual, and ideological understandings of Akan people (Asante and Mazama 2009, 9), Adinkra are associated with death, mourning, and the celebration of life. They are symbolic expressions of funerary messages, believed to be transmitted to us from a departed one comforting the living with symbolic wisdom from the ones that have transitioned. *Sankofa* is represented by a bird standing with its feet facing forward and its head turned backward toward an egg that is resting on its back. It is the egg that sparks my interest in the *study* of this symbol. The bird is not just looking back to take the egg off its back, it is taking care of the egg, making sure to protect its delicate shell. The egg could be seen as a container of the future, and the bird is taking care of the egg as a protector of that future.

To help me conceptualize how a *Sankofa* grounded in a Black feminist ethics of care might look, I have studied the supporting character Nunu in Halie Gerima's 1993 film *Sankofa*. The film presents an understanding of *Sankofa* enacted in the experience of the main character Mona. She is transported from her contemporary life as a model on assignment in and around Cape Coast Castle into the past and the horrors of enslavement on a plantation in Louisiana. Mona, who becomes Shola when she time travels through a dungeon portal in Cape Coast Castle, is enslaved on the Lafayette plantation. In her return to the past, Shola experiences the violent life of the plantation. We see her, branded, raped, whipped and more. At the end of the film, after experiencing the painful visualizing of her circumstance on the plantation, we see Shola sitting looking out to sea, struck dumb by what she experienced. This is one way we could experience *Sankofa*—returning to the past as a constant re-living and suffering of something traumatic and difficult to resolve or reconcile. However, with so much work being done to recognize that this violence happens day in day out, we don't need to time travel to relive the suffering that is happening right now.

Nunu offers us a different way of seeing what *Sankofa* has to offer. Nunu is said to have kept her spirit back in Africa. She was captured from the shores around Cape Coast Castle as a child. Nunu remembers her language and ancient stories, and she shares them with those around her for comfort and hope. She is a midwife, a medicine woman, and the keeper not only of the stories she remembers from Africa but also the stories that speak of the lives of the enslaved. Nunu is an Akan ancestor, a living spirit of *Sankofa*. She is from the matrilineal line that transmits the *mogya*,[25] the blood, the water of life, "the things that are needed to survive" (Asante and Mazama 2009, 49). She offers a way of seeing *Sankofa* differently. She has many ways to escape the tyranny of enslavement, whether that be through her involvement in the revolutionary activities of the enslaved community, through her powers in medicine and conjure or through her ability to transcend space and time through her sleep. She uses her body, her instinct, and her memories to create ways to sustain herself and her community. She shares what she has, making space for collective care and the strength and determination to resist the tyranny of slavery. Of course, it is not always possible to remember languages, ancient stories and movements; loss and separations make it impossible to know the languages or the names of the people that should be close to you. However, within *Declaration of Independence*, the *spiraling undoing of the plot* is a performative embodiment of the bird inviting us to meet our future. Within the collective process we remember together and the things we cannot remember, the languages or the ancient stories come through the sharing of fragments of memory with others that trigger our own rememories. If our grandmothers pass away with their songs, grooming techniques, and knowledges of herbs, we mourn but in the collective process we learn from other womxn, and we find ways to recover that which is lost and imagine other possibilities and ways of being.

Holding Our togetherness

We are on a Zoom call. We are in separate spaces in London, Newcastle, North Wales, Aberdeen, Vancouver, and Grenada. We listen to my friend Anna Marie Gayle as she guides us in a ceremony of movement and togetherness. We are not together, but we are surprisingly close. To feel our closeness, I draw on the strength of the memories of our foremothers buried deep within my body.

Listening to each other
As we dance in the twilight
Holding each other in our hearts
While we stamp our feet on the warm dark red earth
Bodies making ancient gyrations and repetitions
Listening with our eyes
Our hands
Our hips
Hips calling other hips to respond to the movement of the stars
And the crackle of the fire and the beat of the drum
The holler of her voice and the curve of her waist
All my cells are jumping with the energy of this light
Drawing nourishment from the blood vessels that originated in the mitochondria of my ancestors
Finding my soul,
Finding your soul as we dance
They restricted us,
But like ancient times
We found ways to do it anyway
We found profound ways to be together[26]

Figure 8.6 Declaration of Independence, Brent Biennale, December 2020. Photo credit Barby Asante.

To close this chapter, I share a little about the most recent *Declaration of Independence* that was presented online in December 2020 as part of the program of events for Brent Biennale. When the Biennale was presented to me as a possible platform for *Declaration of Independence* in May 2020, it was the early stages of the Covid-19 global pandemic. We had been in lockdown, and, on reflection, there was an overly optimistic hope that it would all be over by the summer. This was not the case, and we had to imagine a *Declaration of Independence* that would happen virtually. Aware of the elements that made *Declaration of Independence*, I had to reimagine how we could generate the sense of togetherness, care, and the feeling of community in four weekly three-hour Zoom call sessions scheduled to devise the performance. I drew on the support of Foluke Taylor, who had been part of the project since 2018 and joined me in Norway, and Black Feminist activist, writer and psychotherapist Gail Lewis, who was involved in the project since the BALTIC iteration. Foluke and Gail assisted me in imagining and creating the space for this iteration as my friends, collaborators and keepers of the spirit of previous declarations. I also invited another friend Anna Marie Gayle to contribute her experience to the vision. Anna runs an occasional movement ritual space for Black and womxn of color called the Black Venus Sanctuary. Using her experience as a 5Rythmns teacher, energy and bodyworker, self-pleasure coach, and theater practitioner, Anna created a powerful virtual somatic ritual of togetherness that called us together despite our separateness. This experience brought release in the form of tears, laughter, and the connection needed to bring a sense of community into a virtual space. It added words to those that Foluke had pulled out of previous performance scripts, to be reflected on and used as prompts for new writing. Gail brought memories of being a young activist supporting the women on the picket line at the Grunwick Film Processing Lab.[27] She reminded us of the ways in which those who came before endured through, a continuing Black feminist anti-racist and anti-capitalist resistance. This resonated with what many of us were feeling at a time when we were witnessing the disproportionate deaths of black and brown people from Covid-related illness, happening because these people with underlying health conditions, were the people laboring in what was defined as essential work while others stayed protected in their homes. We like many others were grieving these deaths and the very public lynching of George Floyd at the hands of US law enforcement that the whole world had witnessed as we stayed home to slow the spread of the virus. Remembering the womxn who came before us, their struggles and their victories, we were nourished by the fact that we could still create and endure despite the conditions.

When it came to the day of the performance, we all gathered in our separate spaces, hoping that the OBS[28] software that professed to make online processes, such as performances, appears smooth with its vison-mixing technology. Sadly, it failed us, but we endured. The broadcast happened 30 minutes later than billed without a single grumble from anybody about the technology.

Ask me again,
Where
I
Come
From
Everywhere
Yes, I declare
Everywhere
we grow, we bloom,
we abound.[29]

Notes

1 Extract from the opening of Declaration of Independence, Feminist Emergency, Birkbeck University of London, June 2017.
2 This passage is from the opening address of *Declaration of Independence* 2017–2020.
3 I use "we" here to reference to those of us who have arrived here as a result of post-war/post-colonial migration.
4 In Ama Ata Aidoo's poem "*As Always, A Painful Declaration of Independence*" (1992), she speaks of the connection between the colonized subject (also woman) and the colonizer as an intimate one. I make reference to this poem's significance to this project later in this text.
5 This passage is from the opening address of *Declaration of Independence* 2017–2020.
6 Projects and campaigns such as Decolonise this Place, Cancel the Damn Art Galleries, and The White Pube highlight this neoliberalization on an institutional level.
7 Extract from the opening of *Declaration of Independence*, all performances, 2017–2020.
8 I use womxn as a term to bring together the complex representation of all those who are and have been part of *Declaration of Independence* cis women, trans women, genderqueer, non-binary, and femme people. It is not ideal and is a contested term.

9 I am thinking about modernity with the decolonial understanding as theorized by Walter Mingolo, drawing on Anibal Quijano's concept of coloniality of power. However, this theory marks the beginning of modernity in 1492 and Columbus's voyage to the Americas. Prior to this time, Europeans were already fortifying Africa. Elmina Castle or Castelo de São Jorge da Mina was erected on the coast in 1482, and Portuguese ships were being sent to the West Coast of Africa by 1418.

10 The Venice Biennale is made up of a central exhibition at the Giardini Della Biennale, official national pavilions in official biennale sites around Venice, and independently organized collateral events recognized by the Biennale Organization. During any Biennale, many other unofficial events, pavilions, and exhibitions also take place.

11 "Propositions #10: Instituting Otherwise," BAK, December 7, 2019, https://www.bakonline.org/program-item/trainings-for-the-not-yet/training-program-trainings-for-the-not-yet/propositions-10-instituting-otherwise/

12 Two episodes were produced in 2021 as a way of reflecting on the process of the project in Norway. https://soundcloud.com/bergen-kunsthall/barby-asante-declaration-of-1; https://soundcloud.com/bergen-kunsthall/barby-asante-declaration-of

13 I'm drawing on an idea of dramaturgy that "is to be found in everything" as proposed by Belgian dramaturge and critic Marianne Van Kerkhoven. She states that a performance's production has a minor zone, "that structural circle, which lies in and around a production" and a major zone in relation to its audiences. She also spoke of the importance of the context of the theater/space and its position, within its city and within the world. Van Kerkoven describes the "walls that link all these circles together are made of skin, they have pores, they breathe" (Georgelou/ Protopapa/Theodoriou 2017: 12).

14 *Declaration of Independence* Brent Biennale 2020 took place online due to Covid-19 restrictions. The surrounding context was an art platform.

15 This is a passage from the opening address of *Declaration of Independence* 2017–2020.

16 I borrow this term from Toni Morrison's *Beloved* (1987). It is used by Sethe to describe not just the things that she remembers or forgets but also things that trigger memories, including some memories that might not be her own but are significant.

17 In *As Always, A Painful Declaration of Independence*, Ama Ata Aidoo spells *undoing* as *on-doing*. This is spelled phonetically as English is spoken by someone whose mother tongue is Fanté.

18 These are three towns in proximity to European forts. There are 32 in total along the coastline of Ghana. Ama Ata Aidoo was born in Saltpond near Fort Amsterdam.

19 1992, 8.

20 1992, 9.

21 1992, 10.

22 From *Declaration of Independence*, Feminist Emergency, Birkbeck College, University of London, and Sonic Soundings Venice, 2017.

23 Dr. Salt is the founder and former co-director of the Centre for Race and Rights at the University of Nottingham.

24 Extract from *Poem for Aba* read as part of *Declaration of Independence*, Opening of the Diaspora Pavilion, Venice, 2017.

25 The blood of life transmitted through the matrilineal line according to Akan lore.

26 Extract from presentation, *Now is the Time for the Performance to Become Reality (I can't do this without you.)*, Propositions #13 Mobilization and (in)Visibility, January 2021.

27 The Grunwick dispute (August 1976–July 1978) for better working conditions and pay was led by South Asian women.

28 Open Broadcaster Software is a free and open software for video and live streaming.

29 Extract from piece by Moira Salt, *Declaration of Independence*, Brent Biennale, December 2020.

References

Aidoo, Ama Ata. 1992. *An Angry Letter in January and Other Poems*. Dangaroo Press, Coventry, UK.

Alexander, M. Jacqui. 2006. *Pedagogies of Crossing: Meditations on Feminism, Sexual Politics, Memory, and the Sacred*. Durham: Duke University Press.

Archive, BALTIC. 2021. "BALTIC Bites: Barby Asante." *BALTIC Centre for Contemporary Arts* Vimeo. Accessed 11 February 2022. vimeo.com/325018686.

Asante, Barby. 2020. "*Declaration of Independence Research*." Declaration of Independence Research and Personal Documentation. doiresearch.tumblr.com/.

Asante, Molefi Kete, and Ama Mazama. 2009. *Encyclopaedia of African Religion*. New York: SAGE.

Class War University. 2012. "Studying Through the Undercommons: Stefano Harney & Fred Moten – Interviewed by Stevphen Shukaitis." Accessed 12 November 12, 2021.classwaru.org/2012/11/12/studying-through-the-undercommons-stefano-harney-fred-moten-interviewed-by-stevphen-shukaitis/.

Dei-Anang, Michael. 1964. *Ghana Resurgent*. Accra: Waterville Publishing House.

Frankétienne. 2014. *Ready to Burst*. Translated by Kaiama L. Glover. New York: Archipelago Books.

Georgelou, Konstantina, et al. 2017. *The Practice of Dramaturgy: Working on Actions in Performance*. Amsterdam: Valiz.

Gerima, Halie. 1993. (Director) *Sankofa* [Film]. Channel Four Films, Diproci, Ghana National Commission on Culture, Mypheduh Films, Norddutscher Rundfunk (NDR), Westdeutscher Rundfunk (WDR).

Harney, Stefano, and Moten, Fred. 2013. *The Undercommons: Fugitive Planning and Black Study*. Brooklyn: Minor Compositions.

Hartman, Saidiya. 2019. "Notes on Feminism: Saidiya Hartman - The Plot of Her Undoing." Feminist Art Coalition. Accessed 11 February 2021. https://static1.squarespace.com/static/5c805bf0d86cc90a02b81cdc/t/5db8b219a910fa05af05dbf4/1572385305368/NotesOnFeminism-2_SaidiyaHartman.pdf.

Imam, Aisha, et al. 2006. "Charter of Feminist Principles for African Feminists." African Women's Development Fund, African Feminists Forum, 15 November 2006.

Kayper-Mensah, Albert W. 1976. *Sankofa: Adinkra Poems*. Accra: Ghana Pub. Corp.

Keeling, Kara. 2007. *The Witch's Flight: The Cinematic, The Black Femme, and The Image of Common Sense*. Durham: Duke University Press.

Gallimore, Olive, Lewis Gail, Quashie Agnes, and Wilson Mabel. 2017. "Black Women Organising: the Brixton Black Women's Group and the Organisation for Women of African and Asian Descent." *Past Tense*, London.

"Gathering: Declaration of Independence." *Brent2020*. Accessed 11 February 2022. www.brent2020.co.uk/whats-happening/programmes/brent-biennial/gathering-declaration-of-independence/.

"Grunwick 1976–1978." *WCML, Working Class Movement Library*. Accessed 11 February 2021. www.wcml.org.uk/contents/protests-politics-and-campaigning-for-change/grunwick/.

"Grunwick, Grunwick Strike 1976–1978." WCML, Working Class Movement Library. Accessed 11 February 2021. www.wcml.org.uk/contents/protests-politics-and-campaigning-for-change/grunwick/.

Lewis, Gail. 2017. "Questions of Presence." *Feminist Review*, 17(1): 1–19.

———. 1996. “Situated Voices: 'Black Women's Experience' and Social Work.” *Feminist Review*, 53(1): 24–56.

“*Library of Performing Rights*.” LADA Live Art Development Agency, 29 November 2019. Accessed 11 February 2022. www.thisisliveart.co.uk/resources/library-of-performing-rights/.

Maja-Pearce, Adewale. 1990. “Ama Ata Aidoo ‘We Were Feminists in Africa First’.” *Index on Censorship* (Women and Censorship), 9: 17–8.

McKittrick, Katherine. 2006. D*emonic Grounds: Black Woman and The Cartographies of Struggle*. Minneapolis: University of Minnesota Press.

Mignolo, Walter, and Catherine E. Walsh. 2018. *On Decoloniality: Concepts, Analytics, Praxis*. Durham: Duke University Press.

Morrison, Toni. 1987. *Beloved*. New York: Vintage Books.

———. 2019. *Mouth Full of Blood: Essays, Speeches, Meditations*. London: Chatto & Windus.

Pambazuka News. “Pan-Africanism, Women’s Emancipation and the Meaning of Socialist Development: Pambazuka News.” Pan-Africanism, Women’s Emancipation, and the Meaning of Socialist Development. Accessed 11 February 2022. www.pambazuka.org/gender-minorities/pan-africanism-women%E2%80%99s-emancipation-and-meaning-socialist-development.

Pranis, Kay. 2005. *The Little Book of Circle Processes: A New/Old Approach to Peace-making*. New York: Good Books.

“Propositions #10: Instituting Otherwise.” BAK. December 7, 2019. Accessed 11 February 2022. https://www.bakonline.org/program-item/trainings-for-the-not-yet/training-program-trainings-for-the-not-yet/propositions-10-instituting-otherwise/

Purkayastha, Madhumita. 2013. “Rememory As A Strategy of Subversive Representation: A Feminist Reading of Morrison’s Beloved Middle Flight. *SSM Journal of English Literature and Culture*.

Rodney, Walter, et al. 2018. *How Europe Underdeveloped Africa*. London: Verso.

Rushdy, Ashraf H. 1992. “Daughters Signifyin(g) History: The Example of Toni Morrison's Beloved.” *American Literature*, 64(3): 567–97.

Sharpe, Christina. 2018. “‘And to Survive.’” *Small Axe: A Caribbean Journal of Criticism*, 22(3): 171–80.

Somé, Sobonfu E. 2006. *Women’s Wisdom from the Heart of Africa*. Colorado: Sounds True Audio Book.

Srilangarajah, Virou. 2018. “We Are Here Because You Were With Us: Remembering A. Sivanandan (1923–2018).” *Ceasefire Magazine*. Accessed 11 February 2022. ceasefiremagazine.co.uk/us-remembering-a-sivanandan-1923-2018/.

Storyvault Films. 2021. “*Declaration of Independence by Barby Asante*.” *Vimeo*. Accessed 11 February 2022. vimeo.com/326049910.

“Striking Women.” The Grunwick Dispute | *Striking Women*. Accessed 11 February 2022. www.striking-women.org/module/striking-out/grunwick-dispute.

Taylor, Keeanga-Yamahtta. 2017. *How We Get Free: Black Feminism and the Combahee River Collective*. Chicago: Haymarket Books.

Temple, Christel N. 2009. “The Emergence of Sankofa Practice in the United States.” *Journal of Black Studies*, 41(1): 127–50.

"The Bird Is Singing Us an Invitation to Meet Our Future: A Workshop with Barby Asante."2019.*ICA*,www.ica.art/live/the-bird-is-singing-us-an-invitation-to-meet-our-future-a-workshop-with-barby-asante.

Thompson, Tonika Sealy, and Stefano Harney. 2018. "Ground Provisions." *Afterall: A Journal of Art, Context and Enquiry*, 45: 120–5.

Weiss, A. Penny. 2018. *Feminist Manifestos: A Global Documentary Reader*. New York: New York University Press.

9 Critical Transnational Queer Praxis

Perspectives on (Re)Production, Performance, and Punishment in the Academy

Andrea N. Baldwin and Alexandra Chandra

Introduction

For decades feminist and queer theorists have written about the limitations of the academy as a space for creativity, radicalism, and transformative work. Chandra Mohanty and M. Jacqui Alexander (2010), for instance, write about the cartographic rules employed to reinscribe the Eurocentric center of the academy even when purporting to address the experiences of people of color. Patricia Hill Collins (1986), in her work on Black feminist thought, also laments that while Black women's theorizing is utilized in the academy, Black women themselves are regulated to the margins. More recently, Roderick A. Ferguson (2012) has stated that "there is a whole school of so-called critical thought and art that despite its oppositional rhetoric, is entirely integrated within the space of consensus" (17) and, Breny Mendoza (2015) has addressed what they refer to as the "diverse technologies of intellectual genocide" (15). In this chapter, we explore the following question: what do the limitations of academia as a place for transformative work mean for those of us passionate about social justice—those of us who are scholar-activists who work, live, and exist in the margins of this space? The implications of the answer (s) to this question have great import for the many strategies currently being utilized or considered for diversifying and refashioning the academy especially at a time when academic enrollments are at a new low, university closures and restructuring due to increased financial pressures are at a new high, political assaults and pressure surrounding issues of free speech and elitism have been stepped-up, and there is the adjunctification of the academy and an assault on tenure. Those who tend to be impacted most negatively by these trends are those already in the margins of academia and hence, we posit, should be implicated in what strategies should be used moving forward.

In this research, we—Andrea N. Baldwin, a Black transnational feminist associate professor and immigrant from the so-called global south, and Alexandra Chandra, a trans-identified woman who is currently working in a technology startup in California after exploring possibilities in the academy—analyze whether the academy, when seen through a critical transnational feminist lens, can be understood (1) not only as a concrete space where social justice work is

DOI: 10.4324/9781003143550-13

stymied due to its inner historical and hierarchal workings but (2) as a space where people bring from their various spatialities and temporalities experience necessary for radical social transformation. Using queer theorist José Muñoz's (1999) concept of disidentification, we examine several aspects of the academic institution and ask whether those of us who are not white male academics employing androcentric epistemologies can successfully work at surviving and transforming academic spaces through what Muñoz refers to as minoritarian performances. Our work also engages queer theory using the lens espoused by Omise'eke Natasha Tinsley (2008); that is, we use "*Queer* in the sense of marking disruption to the violence of normative order ... forging interpersonal connections that counteract imperial desires ... [a] crosscurrent through which to view hybrid, resistant subjectivities – opaquely, not transparently" (199).

We call this framework a *critical transnational queer praxis*. This type of theorizing also follows in the intellectual tradition of Black feminists who ground theory in lived experience (Collins 2000) and who utilize an intersectional lens (Crenshaw 1989). More specifically, the most recent work on percussive feminism developed by Aisha Durham, Brittney Cooper, and Susana Morris (2013) to illuminate "the tension between competing and often contradictory political and cultural projects...that are both discursive and generative" (723–724) was adopted in order to draw attention to the percussive nature of being queer(ed) inside the academy. By holding in place modes of inquiry and ways of being "that might not traditionally fit" (724) and by smashing together queer disidentificatory practices and Black and transnational feminist theory, we create the percussive beat and theoretical pulse of a *critical transnational queer praxis*.

José Esteban Muñoz defines disidentification as survival strategies the minoritized use "in order to negotiate a phobic majoritarian public sphere that continuously elides or punishes the existence of subjects who do not conform to the phantasm of normative citizenship ... [it is a] powerful and seductive site of self-creation" (1999, 4). In academia, minoritized and marginalized folx regularly engage in self-creation and other survival strategies that help us negotiate how our bodies, which are automatically presumed as out of place and are often out of time,[1] (Matthew 2016; Gutierrez y Muhs et al. 2012) become subjected surface planes of signification[2] (Foucault 1972). For some of us who are transnational in our focus, disidentification involves engaging in what transnational feminists refer to as border crossings or cross-border organizing (Boyce Davies 2014; Mindry 2001)—the practice of operating across national, cultural, hierarchical, and other boundaries in order to traverse, transcend and transform them. Baldwin and Johnson's (2018) work on co-mentorship provides a great example of this practice. In their chapter, Baldwin, an adjunct professor, and Johnson, an undergraduate student, write about co-mentorship as a strategy used by black women at a Predominately White Institution (PWI) to resist marginalization. Through co-mentorship, they create for themselves a site of survival in an institution that was not built for them. They write about how crossing hierarchal teacher/student borders was crucial to achieving a counter-space and a new world of sorts where, according to Muñoz, they were

> *not* content merely to survive, but instead to use the stuff of the 'real world' ... and willful enactments of the self for others ... as spectators and actors ... to continue disidentifying with this world until [they] achieve[d] new ones.
>
> (1999, 200)

In this chapter, we use self-reflection to illustrate what a critical transnational queer praxis looks like in our own worldmaking work. Herein we tell our own separate but intertwined stories of survival as a means of leaving behind, like feminists before us, what Grace Kyungwon Hong calls "skeleton architecture" for future generations (2008, 108).

Critical Transnational Queer Praxis: In Theory

In this chapter, we borrow from and build on the critical, active, and self-reflexive theorizing of Black, decolonial and transnational feminists who work systematically and overtly against colonialist, imperial corporatist projects of globalization and capitalist patriarchies. Our approach "challenges mainstream, hegemonic, dominant theories of knowledge, language, power, and politics" (McLaren 2017, 30) while also wrestling with the ways in which individual and collective agency is bound up in these processes (Lock and Nagar 2010). Grappling with these issues means that we are aware of the ways decolonization "has achieved a buzzword status" and has become devoid of meaning when used as metaphor disconnected from "the history of the political decolonization of indigenous territories and lands, and increasingly co-opted by neoliberal market logics" (Hundle 2019, 298), serving to reinvest in ways that actually further settler colonialism (Tuck and Yang 2012). Our approach, therefore, is grounded in engaging complex histories of racial formations of power in order to enact a praxis that is committed to an expansive understanding of decolonization as part of what Hundle refers to as "a much larger archive of anticolonial thought: the study of settler and/or nonsettler colonial projects and the colony itself, studies of colonial mind-sets and attitudes," which includes the academy (2019, 298).

We connect this decolonial work to that of Muñoz's theory of disidentification to develop a percussive praxis specific to the materiality of the minoritized and marginalized at institutions of so-called higher learning to articulate a type of survival strategy which we call a *critical transnational queer praxis*. A critical transnational queer praxis is a heightened awareness of the shared spaces where we reside (whether in the margins or the center), who resides there with us (and who does not), what survival strategies based on our experiences and knowledges we bring from our multiple spatialities and temporalities, and how our survival strategies and worldmaking properties can be beneficial to our existence and future. This praxis requires the minoritized and marginalized to engage in a process of disidentification that is attentive to borders and which is self-reflexive, where we

1 Work toward the breakdown of socially constructed barriers that prevent us from working together and understand that utilizing links engenders change and provides us with the tools to work systematically, collaboratively, and overtly against oppression.
2 Wrestle with the ideological apparatuses disguised as academic disciplines by critiquing the cartographic rules that institutionalize (inter)disciplinary intellectual inquiry.
3 Understand that identities of academic institutions, made consonant with a white heteropatriarchal masculinity, are deeply embedded within a range of reproductive rationales.
4 Use an inherently unstable radical decolonizing praxis, rather than a teleological technique, premised on the notion of being both recursive and self-reflexive to not only situate our intellectual questions outside of present hierarchical systems of knowledge production but also to historicize and share our futures.
5 Interrogate collaboration as a method to critically combine and transform the dialectical relationships between theory and practice, the academic and the activist, the local and the global, the campus and the community, and the center and the periphery.
6 Gesture toward feminist, queer, and other radical genealogies past and present to re-spatialize geographies of power (e.g., by (re)mapping the classroom as a place that reaches beyond the border of the room and transforms the performance of the respectable spectacle into liberatory performances) so that the minoritarian subject can be afforded "the luxury of thinking about the future" and "the privilege or the pleasure of being a historical subject" (Muñoz 1999,189).

We argue that the percussive underpinnings of a critical transnational queer praxis listed above are a minoritarian survival strategy of self-creation. However, this praxis must also be acknowledged for its potential to use the spaces found in academia to indirectly decenter it. As such, while others may see the academic institution as a space that is fraught with structural barriers that continue to marginalize those of us who are not white, male, cis-hetero, able-bodied subjects, we also see academic spaces, such as classrooms, as radical decentering tools (Few, Piercy, and Stremmel 2007; Freire 1972). Our hope is that this percussive praxis will intervene in existing discourse and disrupt discursive fields to illuminate the worldmaking properties found in the survival strategies of marginalized subjectivities in the academy. Like Mariá Lugones (2008), we mean "to begin a conversation … to uncover collaboration … recommit to communal integrity in a liberation direction" (16).

The Androcentric Academy: The Power of the Phallic

Our reflections on the androcentric epistemologies of the academy emerged in response to our own disidentificatory performances inside of it. By (re) producing new worlds through our own continued resistances to academia's rationalized reproductive rubrics, we started "cracking open the code of the majority" in order to expose the "potentiality" of a radical praxis (Muñoz 1999, 31). That is, we explore the possibilities of a critical transnational queer praxis by disidentifying from the traditional logics of the liberal arts. Guided by Muñoz's perspective on minoritarian performances, we began by addressing two questions of concern: (1) What are the conditions inside the US academy that make androcentric epistemologies not only possible but potent? and (2) How have minoritarian identities inside the neoliberal academy produced new transnational and queer cosmologies by disidentifying from rationalized majoritarian reproductive rubrics? For our response, we bring in our own personal experiences to ground our critical praxis in our own worldmaking performances.

In taking this new approach, our theorizing is not only percussive but also palimpsestic as it critiques the imperial bodies of academic and institutional power brokers by building on and borrowing from the work of others who have theorized about the academy. In essence, it creates a new archive/world in which the past/history is still very much present and visible (Spivak 1988). For example, the work of Roderick A. Ferguson, a queer of color critic, is very important to our praxis. Ferguson theorizes about how the rise of many interdisciplinary departments inside the academy was not just a challenge to power but rather constitutive of it and that by incorporating interdisciplinary departments and programs, the academy has in fact brought them within power's control. We also use Chandan Reddy's (2011) model of juridical identity formation, based on categories of difference, to explicitly understand how sexuality gets framed as an amendment to existing modes of juridical recognition that enhance the power of the state and the academy. We strike a beat between Roderick Ferguson's and Chandan Reddy's aforementioned theoretical postulations, Lock Swarr and Nagar's methodological complexity of working through solidarities, Jasbir Puar's (2007) conceptual frame of homonationalism, and Muñoz's perspectives on the performance of politics, to interrogate how the identities of academic institutions, already made consonant with a white heteropatriarchal masculinity, are deeply embedded within a range of reproductive rationales. A critical transnational queer praxis uses a critique of queer pedagogy/performance and transnational feminisms to connect the priapic, white, heterosexual male to curricular colonization and scholarly settler colonialism in order to formulate a larger, less abstracted debate about "cartographies of power" (Alexander and Mohanty 2010). By questioning the who, what, when, where, why, and how of knowledge production and dissemination, this praxis not only challenges the epistemological "eroticisms" and its attendant "masturbatory" research methodologies that are constitutive of this androcentric academy's reproductive rationales, but it also ties together the threads of suffering to open up radical possibilities from the margins. By this we mean that we aim to disrupt discursive fields and to discover worldmaking properties found in the material. This is similar to Anneeth Kaur Hundle's

(2019) project in which they propose that "institutional homes are sites of both estrangements and intimacies … establishing space for alternative presents and futures in the university—new cultures, philosophies, and epistemologies of citizenship, community, and difference that will engender the redefinition of the university itself" (318).

Because our particular praxis lies in the links between the body of the US academy more generally and its androcentrism more specifically, we are interested in the mapping of the masculine and the privileging of the (white) phallic to call into existence the very systems of power that are disguised in their own deployment. This prolegomenous discussion about minoritarian (re)production, performance, and punishment in the neoliberal academy presupposes a larger concern with majoritarian (re)production, performance, and preservation of the neoliberal academy. From this discussion emerges a set of rationalized reproductive rubrics tied to minoritized difference that are controlled through a managerial calculus and are reproduced as sameness through an "add and stir" masturbatory methodology. By a masturbatory methodology, we mean the intentional way in which those in power at institutions intimately perform "socially silenced" acts for their own privately held, coded individualistic pleasure and power that not only bear little relation to critical (pro)creative minoritized performances, but may result in their perceived academic impotence. While the overarching system of the academy positions academic actors in postures to perform for the pleasure of the male gaze, we argue that the androcentric academic institution's autoerotic experiences (i.e., it gets off on seeing what others put out) are often untraceable in historicizing archives. M. Jacqui Alexander (2005) expounds on the colonial concept of "add and stir," where difference is only given cursory attention in a wholly unchanged curriculum (an artificial insemination of sorts). She writes, the

> contours of liberal academic feminism, whose own analytic paralysis in relation to questions of difference, power, class, and racial and sexual privilege gave rise to the 'add and stir feminism' in which the intellectual and political agendas of middle class white women were grounded
>
> (187)

and where women of color, according to Patti Ducan (2014), invoking the metaphor of exoticized Other, find themselves "as both 'hot commodities' … and 'cheap labor'" (41) (see also Nash 2019).

Alexander mentions that "'add-and-stir feminisms' continues to frame the dominant curricular imperatives in far too many women studies classrooms" (2005, 187). In this vein, a number of scholars have already attended to how academic institutions have incorporated minority difference into their structures through interdisciplinary programs and hetero-discursive postures (Ferguson 2012; Reddy 2011). Ferguson argues that "with sexuality's entrance into power's archive, the histories of the gay and lesbian movement were brought into the purviews of institutional consideration, representation, and management" (2012, 209). However, we argue that the will to institutionalize also inaugurated a certain rubric of reproductive repertoires obedient to institutional forms of minoritized difference. We might begin by reading those reproductive repertoires against the white heteropatriarchal identity of academic institutions and against the identity of predominantly white liberal arts institutions deeply invested in libidinal economies of desire. To reproduce is to publish seminal texts in accordance with an "administrative ethos" that artificially inseminates minoritized difference into academic bodies in order to reproduce difference and diversity as hegemonic forms of homogeneity. That is, academic institutions not only neutralize difference according to Ferguson but also neuter it through a range of reproductive rationales aimed at "repressing through its authority…incommensurate publics" (Reddy) and controlling epistemologies that disrupt conventional accounts of knowledge production readily unassimilable inside the androcentric academy.

Examples of how epistemological eroticisms and masturbatory research methodologies combined with the add-stir approach work to erase difference in the academy include: the pressure to publish along a hetero-linear timeline toward a (repro)futurity (Muñoz 2009)[3]; "putting out" publications so the institution can "get off" on and rate faculty's (re)productiveness; being directed to position and contort oneself in "sexy" tenurable postures for a set of punitive student evaluations; performing nothing short of "forced labor" by being forced into labor to reproduce/publish and service majoritarian ideologies for the benefit of a phobic majoritarian public sphere, what Muñoz describes as the burden of liveness.[4] As we know, service or "servicing" inside the neoliberal academic institution is seen as "illegitimate," is perceived as being largely outside of recognized/legitimate institutional apparatuses and neocapitalistic ideologies, and goes mostly unappreciated and unrewarded. Take, for example, the experience Patti Duncan writes about when she states that "women-of-color faculty … have historically been marginalized within academia and expected to perform reproductive labor … for the benefit of others" (2014, 40). Finally, epistemological eroticisms and masturbatory research methodologies can be seen in the privileging of the campus as an exceptional sexualized space in nation-state formation at the expense of camp/us[5] aesthetics. The profound and path-breaking postulation of homonationalism by Jasbir Puar is crucial to our understanding of how the campus may be privileged as an exceptional sexualized space in nation-state formation. By connecting the thread of power between nation-state and the university, between town and gown, we were led to the coining of the term *homolocalism*.

By homolocalism, we mean that the campus and the nation-state are inherently attached politically, culturally, socially, and psychically and equally invested in reproducing majoritarian identity projects. At the small, predominantly white liberal arts institution where we worked and lived, we are aware that the model of community engagement beyond the hill reflects the discourse of American exceptionalism across registers like race, gender, and sexuality. We introduce the term "homolocalism" to build upon Puar's analytic concept of homonationalism in order to draw out an often occluded placental connection between the university and the nation-state. For example, through a concept like homolocalism, we can begin to interrogate what it means for our former institution to be ranked nationally as a top LGBT-friendly campus in the name of diversity, equity, and inclusion. What does it mean to be ranked among other institutions that are vying for the title of "most friendly LGBT campus," and what does it mean to be ranked at all? If, according to Puar, homonationalism is "a facet of modernity and a historical shift marked by the entrance of some homosexual bodies as worthy of protection by nation states" (337), what are the rights accorded to queer bodies protected by the campus as an appendage to or as an amendment of the state? We argue that communities of color and queer communities outside the parameters of the PWI are viewed not only as "hyper-racialized homogenous others" in need of saving but as serviceable (through community outreach programs) figures that stabilize the "queer identity" of the neoliberal academy.

The above analysis highlights the need to develop a strategy that considers not only our positionality but also the social locations of others in and outside of academia, understanding that these locations are inherently unstable and aporetic. Given this, critical transnational queer praxis becomes critical not only to our present survival but to our reproducing a future for ourselves. We ask the questions then as minoritized and marginalized subjects: how are we in our own social locations engaging in a praxis that dismantles hierarchies and barriers through disidentificatory approaches to produce a new world? Are we able to speak our own futurity into being, will it into its existence, or are we foreclosed from such possibilities?

Critical Transnational Queer Praxis: In Practice

The trauma for minoritized folx surviving in a phallic-centric space is well-documented. For example, in *Presumed Incompetent: Intersections of Race and Class for Women in Academia* (Gutierrez y Muhs et al. 2012), women of color in academia report their experiences with microaggression, invisibility, hypervisibility, tokenism, and outright hostility. Our own survival has been a constant topic of discussion and strategizing, alone, together, and with others. Our identities—Alexandra as a queer, biracial, activist, and at the time we met and worked together, staff who engaged in community work; and Andrea as a Black transnational cisgender woman, and at the time visiting faculty in Gender and Women's Studies—have no doubt placed us firmly in academic margins, in similar and different ways. Alexandra navigated monoracial microaggressions and cisheteronormative spaces as a queer mixed-race fem at the time; Andrea navigated a culture of tokenism and racial microaggression. We were both existing in the moment, confined to our own locationalities in academia. But, it is also in this academic labyrinth that we were able to embark upon our own transnational collaboration, "a dynamic construct through which [our] praxis … acquire[d] its meaning and form [was] given place, time and struggle" (Lock and Nagar 2010, 9). We were able to look beyond the commonalities of oppression within the academic system to a praxis of survival and worldmaking.

In the fall of 2016, Andrea was teaching an Introduction to Queer Studies course that Alexandra was auditing. In that course, Andrea had assigned Muñoz's *Disidentifications* as the main text. While discussing the concept of disidentification, Andrea also invited the students into the space as knowledge facilitators/producers and in so doing engaged in a queering of the classroom space by refusing to be bound by the spectacle of classroom etiquette. By asking students to engage in a dismantling of hierarchical barriers and to make their voices audible as knowledge producers in the context of a class about survival strategies for the minoritized, Andrea created a "world" in which we could be recursive, self-reflexive, and engage in liberatory performances (Freire 1972). What follows is our account of how our own self-reflection on our identities and performances as minoritized subjects led us to our own border crossings and to the theorizing of critical transnational queer praxis. We share this experience of enacting a critical transnational queer praxis because a process of decolonizing, as McLaren suggests, "does not remain, nor indeed is it primarily at the theoretical and epistemological level" (2017, 30). In fact, it is an "ongoing struggle for social justice in every sense ... at every level in every arena ... offer[ing] the possibility of new, creative, and innovative approaches to contemporary problems" (2017, 30). We believe that our experiences detailed below provide concrete examples of these struggles and encourage others to think about working themselves into what Kim Tallbear (2018) refers to as "a web of relations ... focusing on actual states of relations – on being in good relation-with, making kin" (161). This type of telling is risky, but risks are important parts of creating our own archives of knowledge as we tell them alongside those with whom we walk. As such, we write our experiences for those at the margins of the academy who are interested in "different ways of inhabiting place, different terms of relationship that chafe against the assumptions" (Vimalassery 2016) of the academy.

Alexandra's Narrative

I am a transwoman, a descendant of diaspora, and a product of transnational border crossings. I was born and raised in Connecticut, in a predominantly white city, by a white mother and an Indian father. I grew up in space, a cultural enclave from which I inherited my mixed-race identity. My childhood home had a lot to do with my thinking about the issues of space and time. My father's "backwardness," that is his relation to a place he always put behind him in his quest for modernity, and my mother's feeling forward made my home an emergent space for an alternative time zone. A great deal of my childhood was also spent listening to my grandparents guide me into their vision of my future. I reflect now on what had been an everyday challenge of calibrating a shared temporality, a temporality that was porous enough to invite multiple temporal connections across a range of spaces. With one foot in each camp and within each time zone, I was close enough to observe and practice the traditions of each race but distant enough to think analytically about them. In this environment, my observation skills were honed to notice those around me. By high school, I was particularly interested in the archetypal images of the universal gay subject reflected around me and sashaying down the hallways, leaning up against the lockers, getting campy in the corridors with wilted wrists, pouty lips, and cocked hips. These images made me critically question my relationship to a community of identities that seemed too stabilized for me. I felt confused about myself during my four years in predominately white hetero high schools, and by the time I graduated, I knew I wanted out.

I attended a PWI for college and then returned to it as a staff member in an office dedicated to community learning and civic engagement—a position I held for a year. During my four years of college, I never once took a course that really spoke to my lived experiences. Most of my courses were taught by white, straight, cisgender professors who theorized from Eurocentric perspectives and engaged in US-based race analysis. At a college that prides itself on its more integrated learning (e.g., when I was working there the college unraveled a four-year educational approach to an interdisciplinary curriculum), it was not elastic enough to deal with the decolonial. My experience felt similar to what Hong (2008) describes as "the university's management of racialized and gendered *bodies* occur[ing] through its management of racialized and gendered *knowledge*" (105). I never took classes that theorized from places of lived experience because the courses I took to satisfy my majors in English and psychology did not offer such theory. To satisfy the requirements for my psychology major, I constantly confronted my own qualms with the discipline. For example, in most of my psychology classes, the limitations section of every research journal we read never really mentioned the primary investigator's identity; this was perhaps a limitation of their own research findings. In fact, speaking from the "I" perspective was wholly prohibited in a discipline that centered the research and not the researcher as ethical grounds for their intellectual projects.

As a student, I also tried pushing the envelope academically but felt pushback from my institution and departments that perceived my research to be methodologically unmanageable. For example, my undergraduate thesis on aging in India took a more transnational turn toward community psychology, the only research in the department that decentered Eurocentric perspectives through phenomenological accounts of aging across borders. As I will soon discuss, Andrea's Introduction to Queer Studies course gave me the radical consciousness I had been craving through its commitment to transnational feminisms and aging. Outside of the classroom, I also found it difficult to find mirrors that reflected aspects of my (dis)embodied experiences. I attended only a couple of my college's LGBT advocacy group meetings until I realized they only spoke to white experiences on campus. After leaving that advocacy group, I started attending queer people of color (QPOC) meetings and found more networks of solidarity there in the margins.

I will forever remember the day I sent Andrea an email kindly inquiring if I could audit her Introduction to Queer Studies course. Although I was no longer a student, I knew I could enroll in one course each semester. I wanted to translate the experiences I was having as a queer mixed-race femme into a larger intellectual project; I wanted to find my politics in a rather apolitcal and anti-intellectual academic milieu. It was in Andrea's class that I began to learn about people of color in the LGBTQIA+ community who have historically been marginalized in our society. Over the course of the semester, I learned about disidentification and "quare" theory (Johnson 2001). I learned about rainbows in the LGBTQIA+ movement refracting only "white" light and about the canonization/commodification of the queer experience. I learned about homonationalism and US exceptionalism.

In tandem with taking her course, I also performed my lived experiences by showing up to class wearing painted nails, necklaces, lipstick, dupattas, and bindis. Taking Andrea's class was extremely imperative in navigating a PWI that continuously punished those whom it asked to perform white, cisgender, heteronormativity. The class guided my subtly subversive performances of transnational belonging by providing the language behind my queer gestures to actively affirm my identity eventually as a transwoman. Donning Indian clothes traditionally worn by women to class made clear to me that I was effortlessly combining energies: shatki (female energy) and shiv (male energy). Relearning history and how media has informed our collective queer imaginary has been extremely important in helping me understand myself. Andrea also actively created an environment in class for students like me to become co-creators of their own knowledge production. For example, I used our course web page discussions to craft an article that was published on an online campaign focusing on queer identities in transnational cultures. I also took my learning from Andrea's class into the community and made connections with local LGBTQIA+ activists and organizers online through media-based activism and attended regularly scheduled support group meetings for queer South Asians. Andrea's class really was a reservoir of potentially utopian energies that helped to foster networks of solidarity. My identity was shaped as much by the folx I encountered outside the classroom as those I encountered in it.

Being connected with folks I identify with has also radically reshaped and restructured my own being and becomingness at home. I am more open now than I have ever been with my family. What I learned in the classroom I was able to translate into grounded application as I passed over the threshold into my house to talk with my parents about my queerness. I gained the confidence, over the course of the semester, to change the narrative of queerness at home by being proximate to those who live and work to uphold it. My family actively affirms my identity as a queer person by asking questions to educate themselves. By comfortably anticipating awkward moments in our conversations, I have grown closer to my family by moving away from established patterns of behaviors and moving toward new unrealized ways of being—making our world new. We recalibrated our previously established understandings of ourselves by engaging in more critical conversations about our own lived experiences that not only resist archival tendencies by embracing the most ephemeral aspects of ourselves but that also bear witness to our own trauma and unbearable realities.

Andrea's Introduction to Queer Studies course also exposed me to the elasticity of queer spatiality and queer temporality, the latter of which I was eager to pursue as a course of study in graduate school. By informing my academic pursuits beyond my time at college and my own familial experiences, I was energized to pursue what I thought, at the time, to be the most intellectually rigorous path possible—a PhD. However, when I took my position outside of academia in 2017, the knowledge I had of academia and the anticipated posturing and potential punishment paled in comparison to the community I found in a space that exemplified the power and potential of the margins. What I found in this space was what bell hooks stated in 1990:

> I was not speaking of marginality one wishes to lose – to give up or surrender as part of moving into the center – but rather of a site one stays in, clings to even, because it nourishes one's capacity to resist. It offers to one the possibility of radical perspective from which to see and create, to imagine alternatives, new worlds.
>
> (149–150)

What I gained in that transnational classroom space on the margins of the academy allowed me to live and thrive in these margins outside of the academy. I am still interested in rethinking existing approaches and methodologies to questions of queer temporality. Specifically, I am interested in queer aging and elders of color and how they embody time (i.e., how their bodies/minds in the present act as a type of archive of the past that resists historicization and yet still somehow gesture toward futurity) and how they are embodied by time (i.e., how the social construction of time is its own site of oppression that triangulates the relationship between race, gender, and sexuality, where those who are not white, young, cisgender, hetero, able-bodied, and masculine, are threatened by the fear of forgetting and no future). I have decided however that I will no longer pursue an academic career. In the margin spaces I have found outside of the academy I get to actively engage with queer elders of color knowledges and have had the privilege to serve these elders instead of researching them. I got to actively live out this transnational queer praxis as a young, middle-class, mixed-raced, trans woman who cross borders as I worked every day with older, mostly working-class, queer folx of color. This was more rewarding for me than being in an environment that prioritizes researching them as subjects over interacting with them as people. Working with Andrea introduced me to what I had been searching for—a critical transnational queer praxis—and helped me become more of what I already am: a sassy, sartorial activist scholar.

Andrea's Narrative

I am a cisgender, heterosexual, Black migrant woman. As a Black transnational feminist whose identity shapes my pedagogy and praxis, I view the academy and spaces in the academy, such as the classroom, as a permeable space where border crossings are possible and where the lines between those who teach and learn, activism and theory, and what constitutes the insides and outsides of the institution/classroom can blur through the use of a type of disidentificatory praxis. A non-hierarchical, collaborative engagement with everyone's experience and interests has the potential to turn the institution/classroom into a less physical and more conceptual space. Such a space allows us to engage in a consciousness raising of sorts which has the potential for awakening temporal disidentificatory activist performances that eventually decenter academia's strict hierarchical divides, thus transforming it into a place with fewer borders. In essence, the institution/classroom transforms into an epistemic gathering place (Pohlhaus Jr. 2017) of potential change so permeable that it becomes indistinguishable as an actual physical space—it becomes a space of worldmaking.

This is the type of classroom space I try to facilitate. As a Black immigrant, first-generation, woman who was at the time I met Alexandra, a visiting assistant professor, such a space was (and still is) important to me as I navigated my tenuousness in academia. To have a messy, semi-non-hierarchal space means that those within it are not expected to perform the burden of liveness. This type of world, while not perfect, creates a reprieve from a very hegemonic white, western, middle-class, heteropatriarchal structure that is academia—one where Black immigrants, like me, find themselves constantly out of place.

While I do not identify as a queer person, as a Black feminist who loves teaching and loves working with my community, my praxis is inherently queer. As Brittney Cooper (2018) writes,

> Black feminism is and has always been a fundamentally queer project … [the] most tried-and-true feminist chicks are open to the possibility of their own queerness because desire is fluid and because boundaries and labels matter so much less when you get down to the real work of what it means to love Black women in the world that hates us all.
>
> (22)

My feminist work is based in an ethic of care for folx who find themselves at the margins. To quote Cooper (2018) again, "one can't truly be a feminist if you don't truly love women [and Others]. And loving women deeply and unapologetically is queer as fuck. It is erotic in the way that Audre Lorde talks about eroticism. It's an opening up, a healing, a seeing and being seen" (20) and loving in the "womanist way that Alice Walker famously talked about – 'sexually and non-sexually'" (21). There is no doubt that my love for my community grounds my politics and my understanding of those in my community as political actors (Harris-Perry 2011, 48).

Black women in academia have written about the difficulty of existing in an academic space that was not made for them (Simmonds 1992; Trinidad 2014; Perlow, Bethea, and Wheeler 2014). Because our intellectual questions, which tend to historicize the lives of our communities as a means of securing our future, are not always seen as situated inside our present system of knowledge production, our work is oftentimes invalidated (Collins 2000). Even for those of us whose work has gained prominence (e.g., Kimberle Crenshaw's Intersectionality), our work becomes misused for androcentric point-making (Lorde 1984); according to Collins, we get gobbled up by a system that appreciates our scholarship over our actual physical selves. As such, Black women who try to exist within academic institutions find that they are stymied by institutional tenure and promotion processes that use criteria like course and department contribution evaluations, which research has shown are more punitive of non-white, non-male faculty (Matthew 2016).

Working then as junior visiting faculty in a small Gender and Women's Studies department placed me in a particularly marginal position. As visiting faculty, I was part of a growing reserve of faculty nationwide who are unprotected by tenure, have little to no job security, low pay, and few benefits (Kingkade 2014). My own marginalization in academia has made me acutely aware of the plight of marginalized students on campus, empathetic to their needs, and thoughtful of how I could make their experience in and outside of the classroom less oppressive. At small liberal arts colleges, this one in particular, the disparities are vast between the haves and the have-nots. First-generation students, low-income students, students of color, and queer students—especially queer students of color who are also international students—tend to be concentrated in the margins. It is these students that I am hopeful to reach through my research, mentorship, teaching, and allyship. Not only students but staff—most of whom are Black and brown—are often left out of the decision-making process and are devalued because they are seen as inherently incapable of producing knowledge and as the help for the good of those who produce knowledge (scheduling their meetings, copying their files, cleaning their offices). It is this awareness of where I am located, where those similarly marginalized are located, and the frustration of the injustice of it all that brought me to the concept of survival through collaboration and border crossing. A practice which uses the stuff of the academic majoritarian world where we find ourselves, to collide with the values and performances of this world in a repercussive fashion, reproducing something new as a means of survival and of building our own space/world within that space—to disidentify.

In so doing, a *critical transnational queer praxis* for me is not only a creative-intellectual project of reimagining what it means to be minoritized and to survive the struggles I encounter in my intellectual life in the academy but which also helps me to rearticulate my own version of who I am through subversive collaborative practices with others who are also located in a place of struggle. As such, I embarked on a number of these projects that center a critical transnational queer praxis by decentering the hierarchical nature of the academic institution. For example, in realizing the importance of identity-based pedagogy that draws from all of our experiences, past and present, I engage, as a professor, in a pedagogy grounded in my transnational feminist ethic of border crossing, which is essential to ensuring my own daily survival and that of marginalized students in the academy. From my academic margins, I continue to cross and destabilize borders in the classroom by incorporating more of my students' identities into my syllabi. I am committed to teaching courses that centered the lives and experiences of the multiply marginalized and analyzing my curriculum for Eurocentric leanings. I also connect with communities of marginalized students on campus to find out what issues are important and appropriate for them to discuss and share in the classroom, while promoting it as a safe space. This process positions the students as knowledge co-producers instead of consumers and leads to several collaborations with students where students were able to articulate their standpoint in and outside the classroom. It also helps students to be and remain engaged, open to others' ideas, excited to conduct research, and invested in knowledge production. I use student suggestions for course readings and content and bring student organizations such as the LGBTQIA+ alliance, Spectrum, QPOC, and others into the classroom to discuss issues of identity-based discrimination and community on campus.

I also listen to what students say are the barriers to their success. At a panel discussion a few years ago focusing on the importance of mentorship for women of color on campus, a Latina student opened up about her job as a campus tour guide for students considering enrolling in college. She shared how her job resulted in her being emotionally conflicted. As a first-generation, lower-income student, she needed to work, but she felt like she was lying to the students of color by presenting the school as an awesome institution while knowing this had not been her experience. Her job provided the financial sustenance she needed and brought her emotional and psychological anguish. The feeling that there was no place on campus where she could turn to for support further compounds this example. There was only one counselor of color on campus, a Black woman, who was already overworked and had a long waitlist. Listening to students helps me to understand where I should focus my energies as an educator and my approach to engaged research.

That student's disclosure made me think more deeply about the various barriers that prevent students from caring for themselves and led to a research project where I collaborate with students to examine how the neoliberal academy with its superficial rhetoric and performances of equity and inclusion prevents students from engaging in individual and collective care practices. From this project, I was able to produce a short documentary with a former student—Daryl Brown. This film has since been screened at The Berlin Feminist Film Festival in 2019.

This collaborative work also extended to staff who are also often marginalized on university and college campuses. My work with Alexandra began when she asked to audit my Introduction to Queer Studies course in the fall of 2016.

I teach this course from the perspective of QPOC. It is no secret that QPOC theorizing is not widely taught in higher education classrooms and that mainstream queer studies silences the voices of QPOC (Bravmann 1997). Committed to my own transnational praxis as a non-queer-identified person, I made a decision to make connections with QPOC, to make their stories audible in my pedagogy, and to bring these marginal voices into the classroom. I use the work of José Esteban Muñoz, Gloria Anzaldua, Cherrie Moraga, Cathy Cohen, Marlon Ross, Marlon Riggs, Suk Han, E Patrick Johnson, Audre Lorde, and Barbara Smith, just to name a few. I also draw on my transnational relationships with QPOC and introduce my students to queer folx from the global south as people who are also constrained by the legacy of colonialism and imperialism, who are agents for their own cause, and who use both cultural traditional and innovative strategies to fight oppression.

It is through opening up my classroom to a person who as staff is seen as out of place that Alexandra and I were able to embark upon an intellectual project, which built on her identity (then as a queer, mixed-raced femme) and my own transnational identity and queer praxis as a Black feminist scholar. What makes our work queer and transnational is not only our identities, Alexandra's performance, or Andrea's commitment to queer allyship but rather the disidentificatory elements of our collaborations and commitment to queer life in academia. Our separate cross-border disidentificatory journeys, our act of professor-staff collaboration, and the sharing of ourselves so openly and honestly with the purpose of securing our own survival have resulted in the creation of this percussive theorizing. Inherent in our critical transnational queer praxis is what Muñoz refers to when he writes about worldmaking.

The Worldmaking of Critical Transnational Queer Praxis

The crux of making a new world is bound up in the six percussive underpinnings outlined earlier in this paper. When Andrea welcomed Alexandra into her classroom to audit Introduction to Queer Studies and then again to share her experiences as a knowledge facilitator in a subsequent class, she was engendering change by utilizing the spaces the academy provides, working systematically, collaboratively, and overtly against oppression by positioning/presenting Alexandra's knowledge as valuable and knowable.

This praxis refutes the ideological apparatuses disguised as academic disciplines by critiquing the cartographic rules that institutionalize (inter)disciplinary intellectual inquiry. This critique happens through the use of one's lived experience as a place from where one can teach, learn and theorize but not by using "colonizing" techniques in the classroom, by trying to give a language/academic speak to the "subaltern" (Spivak) or expecting them to regurgitate the jargon of a particular discipline. Rather, by Andrea sharing her own life stories and engaging with students as co-knowledge producers and facilitators, this praxis affords students the opportunity to speak their own truth to power and a space in which to understand the world in the context of their own lived experiences. For example, Alexandra was able to create for herself a new world through this engagement. Different from her undergraduate experience that offered her a career path, this experience provided Alexandra guidance that changed her future trajectory, revealing to her more about the world that she wanted to live in and her calling.

As this praxis made it clearer to us how academic institutions are deeply embedded within a range of white, male reproductive rationales, as detailed earlier, we saw that what we were doing had to be about more than survival. We needed to understand and claim ourselves as historical subjects as a means of making certain that our daily work could ensure the possibility of a future for those of us in the margins. According to Hong (2008), this "does not concede the future to the present, but imagines it as something still in the balance, something that can be fought over … the work … of imagining, is revolutionary" (108). With these goals in mind, we began to more deliberately interrogate our work together as more than once-in-a-while collaborations or invitations to facilitate a class but as a method to critically combine and radically change the dialectical relationships between theory and practice; the academic, staff, and the activist; the local and the global; the campus and the community; the center and the periphery. We sought to enact what Gaile Pohlhaus Jr. (2017) refers to as "epistemic gathering," work that "develops new ways of acting in concert" (98). According to Pohlhaus Jr., this gathering does not happen all at once but comes with time, practice, commitment, and struggle as we engage in worldmaking "making intelligible through our embodied epistemic practices that which dominant epistemic institutions continually and relentlessly render unintelligible in their own commitments to serve dominant interests" (98–99). It is through our experience being in the community "[m]aking sense from and with (rather than just of) these experiences [that]… allow[ed] such experiences to make an impression that moves us, epistemically speaking" (Pohlaus Jr., 99).

This was how the world of critical transnational queer praxis was born. The stuff of this world is feminist, and it is queer. It "doesn't merely reflect the material world, nor is it an epiphenomenon of it, but rather is a 'skeleton architecture,'" (Hong, 108) the scaffolding for future worldmaking. It is akin to the way in which Black lesbian feminist poet Audre Lorde situates poetry "as the *base*" (Hong, 108), one from which Black queer feminist such as Alexis Pauline Gumbs is currently creating and which lives in the present reimagining and reworking the academic experience as a "less disciplining definition of knowledge" (Hong, 108). Gumbs self-defines as "a Queer Black Troublemaker and Black Feminist Love Evangelist and an aspirational cousin to all sentient beings … [whose] work … is to facilitate infinite, unstoppable ancestral love in practice … [and is] in response to the needs of her cherished communities" (https://www.alexispauline.com/about). Gumbs' work is connected to ours in that "web of relations" mentioned by Tallbear, as it asks, for example, what does it mean "to be nobody in a university economy designed to produce somebody individuated, assimilated, and consenting to empire" (Gumbs 2014, 237)? Gumbs "offers… the idea of a larger 'we,' created by the truly poetic"

(240), that is the "practice of developing a relationship with others that is not about negating the other to produce a human self but rather about the human poetry of creating and describing possible collective relationships to the environment" (241). If nothing else, we hope to expand this web of relations as part of a "skeleton architecture," a palimpsest for future worldmaking similar to the genealogical legacy from which we currently build and write upon.

Our praxis pulls and gestures toward other radical genealogies past and present (as Andrea pulls from her Caribbean roots and Alexandra her bi-racial identity) to re-spatialize geographies of power to create spaces in the academy that reach beyond the border of the room. To create spaces that transform the performance of the respectable spectacle into liberatory performances so that the minoritized can be afforded "the luxury of thinking about the future" (Muñoz, 189) and "the privilege or the pleasure of being a historical subject" (Muñoz, 189). We join in the legacy of Lorde and Gumbs, viewing the "public space of the classroom and the production of [something like] poetry for the multiple generations [,] creat[ing] a temporal intervention … [t]he poetics of survival, a queer relationality… teaching, an undisciplining impulse in the classroom, a counterproductive mode … demonstrating … solidarity" (Gumbs 254). This palimpsestic work is imperative as we "take that work further, securing a future for those who come after" (Baldwin 2020, 154).

Conclusion

As we continue to refine our work on critical transnational queer praxis, we do so in the context of our lived realities where Alexandra works in the tech industry and Andrea has since transitioned to a larger public institution and is now tenured. We both have found that working together caused the sort of intellectual friction we needed to create a new world in the one we lived and also in the ones we now inhabit. It has been through Andrea's praxis of re-spatializing her classroom that we have engaged in a further remapping of other spaces—for example, late night meetings at dining halls around campus, phone and text conversations, conference planning in airports, rendezvous at Andrea's home, and voguing our presentations outside of the US nation-state's socially constructed borders—that we have found the perfect algorithm for a disidentificatory project that finally maps out the affective flows between those the androcentric academy works to keep apart: a Black immigrant woman faculty and a trans staff member.

Notes

1 By out of time we are referring to the demands of the tenure clock. When it comes to issues of promotion and the tenure, statistics indicate that academics of color are most likely to be denied tenure once it is time to go up for promotion and are also often not on the tenure clock because they are over represented in non-tenure track ranks.

2 Michael Foucault's philosophical concept of subjectification is useful here to our own understanding of the construction of the subject inside the neoliberal academy.

3 In *Cruising Utopia: The Then and There of Queer Futurity*, Muñoz (2009) describes futurity as "not yet here." We preface futurity with (repro) to emphasize that to get t/here, that is to get tenure, one is expected to reproduce in very specific Eurocentric positivist ways. We also gesture to Edelman's (2004) concept of reproductive futurism, a process where political decisions are made on behalf of the unborn child. In the same vein, political decisions in academia have supposedly been made in our (minoritarian subjects) best interest even before we enter the space. In fact, there is "an ideological limit on political discourse… preserving in the process the absolute privilege of heteronormativity by rendering unthinkable, by casting outside the political domain, the possibility of a queer resistance to this organizing principle of communal relations". By organizing in this way, academia historically excludes those in the margins, including queer folx and people of color since we are not seen as capable of reproduction at all or not able to reproduce in the same ways.

4 Muñoz describes the burden of liveliness as an "inflection of the experience of postcolonial, queer, and other minoritarian subjects" to perform for the amusement of the dominant power bloc as "the only imaginable mode of survival" (187).

5 We separate "camp" and the pronoun "us" here to think about how queer utopian energies are animated perhaps not only by a "we" as has been theorized extensively but also by an "us" that incorporates its orthogonal other "them".

References

Alexander, M Jacqui. 2005. *Pedagogies of Crossing: Meditations on Feminism, Sexual Politics, Memory, and the Sacred.* Durham: Duke University Press.

Alexander, M Jacqui, and Chandra Talpade Mohanty. 2010. "Cartographies of Knowledge and Power: Transnational Feminism as Radical Praxis." In *Critical Transformational Feminist Praxis*, eds. Amanda Lock Swarr and Richa Nagar, 23–45. Albany, NY: State University of New York Press.

Baldwin, Andrea N. 2020. "Lugones, Munóz, and the Radical Potential of (Dis)identificatory Feminist Love for "World"-Making Beyond the Academe." *Frontiers: A Journal of Women Studies*, 41(1): 141–60.

Baldwin, Andrea, and Raven Johnson. 2018. "Black Women's Co-Mentoring Relationships as Resistance to Marginalization at a PWI." In *Black Women's Liberatory Pedagogies: Resistance, Transformation, and Healing Within and Beyond the Academy*, eds. Olivia N. Perlow, Durene I. Wheeler, Sharon L. Bethea, and Barbara M. Scott, 1st ed., 125–40. New York: Palgrave MacMillan.

Boyce Davies, Carole. 2014. "Pan-Americanism, Transnational Black Feminism and the Limits of Culturalist Analyses in African Gender Discourse." *Feminist Africa*, 19: 78–93.

Bravmann, Scott. 1997. *Queer Fictions of the Past: History, Culture, and Difference*. Cambridge: Cambridge University Press.

Collins, Patricia Hill. 2000. *Black Feminist Thought: Knowledge, Consciousness, and the Politics of Empowerment*, 2nd ed. New York, NY: Routledge.

———. 1986. "Learning From the Outsider Within: The Sociological Significance of Black Feminist Thought." *Social Problems*, 33(6): 14–32.

Cooper, Brittney. 2018. *Eloquent Rage: A Black Feminist Discovers Her Superpower*. New York: St. Martin's Press.

Crenshaw, Kimberle. 1989. "Demarginalizing the Intersection of Race and Sex: A Black Feminist Critique of Antidiscrimination Doctrine, Feminist Theory and Antiracist Politics." *University of Chicago Legal Forum*, 1989(1): 139–67.

Duncan, Patti. 2014. "Hot Commodities, Cheap Labor: Women of Color in the Academy." *Frontiers: A Journal of Women Studies*, 35(3): 39–63.
Durham, Aisha, Brittney C. Cooper, and Susana M. Morris. 2013. "The Stage Hip-Hop Feminism Built: A New Directions Essay." *Signs*, 38(3): 721–37.
Edelman, Lee. 2004. *No Future: Queer Theory and the Death Drive*. Durham: Duke University Press.
Ferguson, Roderick A. 2012. *The Reorder of Things: The University and Its Pedagogies of Minority Difference (Difference Incorporated)*. 1st ed. Minneapolis: University of Minnesota Press.
Few, April L., Fred P. Piercy, and Andrew Stremmel. 2007. "Balancing the Passion for Activism with the Demands of Tenure: One Professional's Story from Three Perspectives." *NWSA Journal* 19(3): 47–66.
Foucault, Michel. 1972. *The Archaeology of Knowledge: And the Discourse on Language*. New York: Pantheon Books.
Freire, Paulo. 1972. *Pedagogy of the Oppressed*. New York: Herder and Herder.
Gumbs, Alexis Pauline. 2014. "Nobody Mean More: Black Feminist Pedagogy and Solidarity." In *The Imperial University: Academic Repression and Scholarly Dissent*, eds. Piya Chatterjee and Sunaina Maira, 237–59. Minneapolis: University of Minnesota Press.
———. "About." Accessed 21 June 2020. https://www.alexispauline.com/about.
Gutierrez, Y. Muhs Gabriella, Yolanda Flores Niemann, Carmen G. Gonzalez, and Angela P. Harris. 2012. *Presumed Incompetent: Intersections of Race and Class for Women in Academia*. Boulder, CO: University Press of Colorado.
Harris-Perry, Melissa. 2011. *Sister Citizen: Shame, Stereotypes, and Black Women in America*. New Haven: Yale University Press
Hong, Grace Kyungwon. 2008. "The Future of Our Worlds: Black Feminism and the Politics of Knowledge in the University under Globalization." *Meridians: Feminism, Race, Transnationalism*, 8(2): 95–115.
hooks, bell. 1990. *Yearning: Race, Gender, and Cultural Politics*. Boston: South End Press.
Hundle, Anneeth Kaur. 2019. "Decolonizing Diversity: Transnational Politics of Minority Racial Difference." *Public Culture*, 3(12): 289–322.
Johnson, Patrick. 2001. "'Quare' Studies OR (Almost) Everything I Know About Queer Studies I Learned from my Grandmother." *Test and Performance*, 21(1): 1–25.
Kingkade, Tyler. 2014. "9 Reasons Why Being an Adjunct Faculty Member is Terrible." The Huffington Post. Accessed 12 August 2018. http://www.huffingtonpost.com/tyler-kingkade/.
Lock, Swarr Amanda, and Richa Nagar. 2010. "Introduction: Theorizing Transnational Feminist Praxis." In *Critical Transformational Feminist Praxis*, eds. Amanda Lock Swarr and Richa Nagar, 1–20. Albany, NY: State University of New York Press.
Lorde, Audre. 1984. *Sister Outsider*. New York, NY: Random House Inc
Lugones, Mariá. 2008. "The Coloniality of Gender." *Worlds & Knowledges*, 2: 1–17.
Matthew, Patricia A. 2016. *Written/Unwritten: Diversity and the Hidden Truths of Tenure*. Chapel Hill: The University of North Carolina Press.
McLaren, Margaret A. 2017. "Introduction: Decolonizing Feminism." In *Decolonizing Feminism Transnational Feminism and Globalization*, ed. Margaret A. McLaren, 17–27. London: Rowan and Littlefield International.
Mendoza, Breny. 2015. "Coloniality of Gender and Power: From Postcoloniality to Decoloniality." In *The Oxford Handbook of Feminist Theory*, eds. Lisa Disch and Mary Hawkesworth, 2–26. Oxford: Oxford.

Mindry, Deborah. 2001. "Nongovernmental Organizations, 'Grassroots,' and the Politics of Virtue." *Signs*, 26(4): 1187–211.
Muñoz, José. 2009. *Cruising Utopia: The Then and There of Queer Futurity*. New York: NYU Press.
———. 1999. *Disidentifications: Queers of Color and the Performance of Politics*. Minneapolis: University of Minnesota Press.
Nash, Jennifer C. 2019. *Black Feminism Reimagined After Intersectionality*. Durham: Duke University Press.
Simmonds, Felly Nkweto. 1992. "Difference, Power and Knowledge: Black Women in Academia." In *Working Out: New Directions for Women Studies*, eds. Hilary Hinds, Ann Phoenix, and Jackie Stacey, 52–60. London: Falmer Press.
Perlow, Olivia N., Sharon L. Bethea, and Durene I. Wheeler. 2014. "Dismantling the Master's House: Black Women Faculty Challenging White Privilege/Supremacy in the College Classroom." *Understanding & Dismantling Privilege*, IV(2): 242–59.
Pohlhaus Jr., Gaile. 2017. "Knowing without Borders and the Work of Epistemic Gathering". In *Decolonizing Feminism Transnational Feminism and Globalization*, ed. Margaret A. McLaren, 37–54. London: Rowan and Littlefield International.
Puar, Jasbir. 2007. *Terrorist Assemblages: Homonationalism in Queer Times*. Durham: Duke University Press.
Reddy, Chandan. 2011. *Freedom with Violence: Race, Sexuality, and the US State*. Durham: Duke University Press.
Spivak, Gayatri C. 1988. "Can the subaltern speak?." In *Marxism and the Interpretation of Culture*, eds. Cary Nelson and Lawrence Grossberg, 271–313. Champaign: University of Illinois Press.
Tallbear, Kim. 2018. "Making Love and Relations Beyond Settler Sex and Family." In *Making Kin Not Population*, eds. Adele Clarke and Donna Haraway, 145–63. Chicago: Prickly Paradigm Press.
Tinsley, Omise'eke Natasha. 2008. "Black Atlantic, Queer Atlantic: Queer Imaginings of the Middle Passage." *GLQ: Journal of Lesbian and Gay Studies*, 14(2–3): 191–215.
Trinidad, Alma M.O. 2014. "The Becoming of a Pinay Scholar Warrior of Aloha: A Critical Autoethnography of Teaching, Mentoring, and Researching for Social Change. *Polymath: An Interdisciplinary Arts and Science Journal*, 4(2): 17–38.
Tuck, Eve, and K. Wayne Yang. 2012. "Decolonization is not a metaphor". *Decolonization: Indigeneity, Education & Society*, 1(1): 1–40.
Vimalassery, Manu. 2016. "Fugitive Decolonization" *Theory & Event*, 19(4). Accessed 21 June 2020. 2018. https://muse.jhu.edu/article/633284.

Section IV

Black Feminisms and Healing Futures

10 'Tacit Sexualities' Transforming the Narrative

Afro-Caribbean Women and the Politics of the Body

Evette Burke

Introduction

Viviana Zelizer, in her book *The Purchase of Intimacy* (2005), cautions against narrowly defined categorizations of sex and economic relations and the futility of constructing intimacy[1] and economics as separate, mutually exclusive spheres. Zelizer proposes that the complexities of intimacy and economic relations are peculiar to the extent that they preclude rigid categorizations. Likewise, Barry Chevannes recognizes that the multi-layered attributes of sex and economic relations allow for emotional, material, and practical applications (2001, 16). Yet despite the acknowledged fluidity of sex and economic relations, the politics of sexuality has through regulatory mechanisms, legitimized some forms of sexual conduct, denigrated some sexual acts, and identified as problematic othered sexual bodies.

Black feminists' concerns with Western relations of domination have exposed the intentionality of colonial narratives to occasion the devaluation of black womanhood (Collins 2000; hooks 1981a). Evelynn Hammonds writes of White European elitist construction of the Black woman's sexual body as diseased, deviant, and associated with prostitution (2008). Upholding the devaluation of Black women's bodies is the Western contemporary construct of Black women's sexual embodied practice, deployed through the label '*transactional sex,*' as *the exchange of sex for money or goods*, (Leclerc-Madlala 2003; Robinson and Yeh 2011; Swidler 2007). This label is especially bestowed on sub-Saharan African and to some extent, Black Caribbean women.[2] As presently constituted, the term 'transactional sex' and its discursive association with prostitution, Black bodies, and disease uphold the notion of problematic sexuality. This association serves as a reminder that the stereotypical configurations of Black women's sexuality have not shifted.

Ann Swidler and Susan Watkins observe that the term 'transactional sex' and its attendant meanings were constructed by Western observers and forced on the bodies of poor African women (2007). However, hegemonic relations 'naming' Black women's bodies as transactionally sexualized commodities do not foreclose individual or collective agency. We are reminded that 'self-definition' enables Black women's resistance against racialized claims about our collective existence (Collins 2000, 10).

DOI: 10.4324/9781003143550-15

The work produced in this chapter derives from a broader exploration of intimacy and economic relations of Afro-Guyanese[3] women. And though it considers women in the Anglophone Caribbean generally, it speaks particularly to Afro-Caribbean women. Shaped by feminists' epistemological and methodical approaches, this work supports two important considerations for Black feminists theorizing. The first is; that it matters who speaks for us. Rowley (2000) endorses this consideration by acknowledging the power of the 'voice' in naming one's subjective experience. The second consideration heeds Black feminists' call to transform hegemonic narratives which sustain racialized tropes by redefining knowledge of ourselves. Therefore, this work is understood within the context of a re-definition.

What you will read here is not a study of 'transactional sex.' I do not seek to compare young people's sexual practices defined as 'transactional sex' on the African continent, with those described as 'transactional sex' in Caribbean literature. Neither do I seek to redefine the term 'transactional sex' nor to script 'transactional sex' on Black women's bodies. I argue against locating young Afro-Caribbean women's non-marital heterosexual relations of intimacy and economics within a synonym for prostitution. I offer a new lens for theorizing Afro-Caribbean women's sexual relations that dislocate their sexual bodies from the controversial reading of 'transactional sex.'

To sustain this position, I examine; (a) young Black women's challenges with race, gender, and sexuality, and how their sexuality is deployed within the effects of power; and (b) the attributes, practices, and complexities of their intimate and economic relations including their ruptures from or association with other relations involving sex and money.[4] This examination produces an assemblage of structuring pillars of resources and relations. I describe that assemblage as *'tacit sexualities,'* which I define as: non-marital, non-cohabiting, sexual relations, mainly between single people in which sex, functions in coterminous agreement/value with emotional investment and financial social support, to sustain the relationship. Their sexual activity is not considered labor or prostitution, by those who engage in these relations nor by those in the society. There is no client/patron or formal contractual arrangement. There is no fixed or predetermined payment, no fixed or negotiated value for sexual activity. There is no agreement of immediate compensation or promise of compensation with each sexual encounter. These relations are largely monogamous for one or both of the parties. During the relationship both or one of the parties, consider her/himself to be in love. In 'tacit sexual' relations, the affective attachment creates increased familiarity which allows for the performance of non-sexual activities. These relations of high social tolerance tend to be fractious because of the often-unstable commitment to love, sex, and money on which they rest.

The concept 'tacit' – to which I shall return later – explains the unspoken or implicit manner in which many Afro-Caribbean single women navigate the monetary or material transfers from men with whom they have sex. In my work, 'tacit' connotes a 'silent knowing' that produces particular material and sexual, outcomes in many Afro-Caribbean women's relations. And though love, sex, and money are simultaneously required for a viable intimate relationship, money and sex remain the 'tacit' aspects of their sexualities. Ulises Decena writes that 'A tacit *subject* might be an assumed and understood, but not spoken, aspect of someone's *subjectivity* as well as a particular theme or topic' (2011, 31). Unlike Decena (2011, 38) who deploys tacit subjects as 'more a point of departure than a frame for the overall discussion,' I develop tacit sexualities as a frame of analysis.

Hegemonic Discourse and the Naming of 'Problematic' Bodies

The term 'transactional sex' originated around epidemiological concerns about commercial sex workers and HIV in the mid-1990s. Contemporary readings on the term reflect a geopolitical shift to Africa and the Anglophone Caribbean at the intersections of youth, place, gender, sexuality, and race. The paradigms linking transactional sex with HIV invite concerns of 'unregulated' Black sexual bodies, endemic HIV, sexual materiality, risky behavior, and irresponsible sexual practices (Maganja 2007; Stoebenau 2011; Wamoyi 2010; Luke 2002; Hunter 2002; Barrow 2008).

Several United Nations (UN) documents affirm the nexus between transactional sex and prostitution. Whether as a reminder to UN members about the multiple practices and shifts in patterns of sex work, including transactional sex (UNAIDS 2009), in defining a 'sex worker' as 'any woman who engages in transactional sex,' (Robinson 2011, 36) in discussions on 'informal' sex workers who supply transactional sex less frequently (3), or sharing a numerical quantification that 'women engaging in transactional sex in 25 European countries suggest a sex worker population of some 700,000 women' (Noble 2010, 49).

Jonathan Robinson and Ethan Yeh's 2011 study typifies the common discursive categorization of transactional sex; they claim:

> Exchanging sex for money goods or services is a way of life for many poor women in developing countries, yet little is understood about the way that the commercial or transactional sex market functions. While commercial sex workers have long been identified as critical in affecting the spread of the HIV/AIDS epidemic, comparatively little work has gone beyond characterizing sex workers as a high-risk population.
>
> (35)

This excerpt reinforces three descriptive characteristics of transactional sex in the literature: (a) exchange of sex for money or goods, (b) commercial sex work, and (c) sex market relations intrinsic to sex and commerce. Though Robinson and Yeh connect transactional sex to poverty, the UN's intentional use legitimizes the term and makes valid its application within frameworks of sexuality as dangerous. In the Caribbean and other parts of the world, society responds to 'dangerous sexuality' by moral condemnation, social marginalization, and the lawful regulating and policing of the sexual body.

The broad investigative interest in transactional sex and young African women produces several notable themes. Stark (2013) addresses pure sex and material exchanges between girls/ women and men. Kristen Stoebenau et al. consider transactional sex, HIV, and risky behavior (2011); R.K Maganja examines transactional sex, HIV, and poverty in Africa (2007) and Ann Swidler considers sex relations of interdependence (2007). The literature not only positions African bodies in transactional sex; it distinguishes between the sex and economic practices of North American and European women and Black women. Hope's (2007) desk review considers the vulnerabilities and risky behaviors of young African and Jamaican women in transactional cross-generational sex relations. For her, these young people forgo concerns of safety and risk in favor of immediate benefits. With specific reference to US teens with older partners, Hope asserts: 'U.S. teens with partners six or more years older are nearly four times more likely to become pregnant' (16). Hope avoids naming those relations as 'transactional.' And does not appear to attribute the absence of condom use to risky behaviors of US teens, which invariably is associated with pregnancy.

Correspondingly, Carol Kaufman and Stavros Stavrou studied sex and gifts in the relations of white and Indian suburban youth, and Black youth from townships, and the inner city, of Durban, South Africa (Kaufman and Stavrou 2004). They revealed white and Indian women's resistance to foregoing condom use if a gift were offered for sex. Conversely, if African women 'wanted' to see the man again, then condom use becomes a non-issue and they would take on the risk (387). Arguably, risk-taking is a prerogative of youth. Nevertheless, class distinctions among the groups – as their sample suggests – coupled with the socio-economic deprivation of those from townships may force many poor women to make survival choices based on the resources available to them. Swader's (2013) study of Russian women's engagement with sex for material gains detaches those relations from 'prostitution' or 'transactional sex.' Rather, they categorized these practices as 'Gift-for-Sex Barters,' (GFS) barters, or 'compensated dating' (2). Swader argues that (GFS) barters are incomparable to prostitution or 'transactional sex' as written in studies of African women because of the absence of money. Swader claims that the 'barter' arrangements where sex is exchanged for cars, fur coats, and iPhones hinder exact calculations of what is exchanged, and its flexibility allows women to postpone sex. In addition, it accommodates greater emotional involvement and reduces the power dynamics, unlike that involving prostitution. In (GFS) barters, Russian women seek older, generous, accomplished married or unmarried men, and men seek young, pretty women for periodic sexual encounters (2013).

Swader's distinction between (GFS) barters, transactional sex, and prostitution is at best tenuous. The 'barter' process allows for the negotiation of 'wants' that morally if not contractually, obligate both parties to deliver on the terms agreed. Evident in this process are features of the 'transactional' as in prostitution. And though 'barter' complicates the exact calculation of sex to commodities exchanged, it cannot prevent the negotiating parties from considering their exchanges in monetary value. Thus, to differentiate between (GFS) barters, prostitution, and transactional sex, based on the absence of money, is an attempt to obscure the commodified nature of the relations. Factually, the women in these relations are not exchanging sex for affection; rather, they 'barter' sex for luxury consumption goods.

Transactional Sex in Caribbean Literature

In the Anglophone Caribbean, the focus on issues of youth sexuality and reproductive health has now come to consider relations of transactional sex (Phillips 2006; Kempadoo 2006; Barrow 2008). Christine Barrow's accounts of the 'bashment' youth experience discuss 'transactional relationships' where girls' exchange of sex for commodities is fundamental (2008, 8). Daphne Phillips correspondingly, perceives that the 'operations of the market' of sex and material exchanges now proliferate in young people's sexual behavior, in Montserrat (2006, 52). Curtis' (2009) study of young women in Nevis asserts that sex and monetary exchanges sit within the organizing principle of young Caribbean women's sexual life, and Kamala Kempadoo claims that sex in the Caribbean is often transactional (2006). In fact, she heavily associates transactional sex with young Caribbean people (2011). Curtis however adds the dimensions of place, race, and sexuality by claiming a point of sexual difference between 'commodity erotics' or 'transactional sex' and her Western orientation that connects sexuality with love (2009, 17). Curtis shapes her notion of sexual difference by her belief in the myth of Black women's sexual immorality. At the same time, she chooses to forget that eighteenth-century Western ideals of romantic love first considered the socio-economic status of the person desired (Illouz 2006).

The readings on transactional sex challenge one to distinguish the act from prostitution due to the similarity in characteristics, motivation, and practices. The term also lacks definitional clarity because of its conflation with prostitution, while simultaneously generating conceptual ambiguities for other sexual relations involving money. Further, the failure to explain the formation of transactional sex exchanges clouds how one may filter theoretically the different categories of practices involving sex and money.

Transactional sex, when codified to elide categorizations outside the framework of prostitution, is inherently laden with power, interest, and value. Created from the lexicon of hegemonic power, its discursive formation sets the mode that marks the lives of those occupying selected geopolitical spaces. Michel Foucault's reading of discourses as sites of power and resistance producing and organizing rules and meaning is instructive (1972, 1990). Not only for understanding the power relations 'naming' and 'defining' the contexts of Black women's subjective experiences but for resisting productive power as a mechanism of sexual control. Evident in all this is a politics that produces Black women's visibility through their overexposure to transactional sex and their simultaneous invisibility by failing to sufficiently acknowledge young

Black women as sexually healthy bodies seeking sexual pleasure. The failure to comprehensively differentiate between young people's relations of mere sex exchanged deliberately for money or goods and money or goods provided when relations are sexual is troubling. This failure further triggers concerns about simplifying or ignoring a very complex set of arrangements in young Afro-Caribbean women's sexual relations. Factually, Afro-Caribbean women's culture and experiences hold greater relevance for a re-thinking of the ways theories and concepts are constructed and imposed on their bodies.

Theoretical and Methodological Underpinnings

Re-thinking the scripting of transactions sex on young Caribbean women's bodies required an examination of the intimate and economic practice of Afro-Guyanese women. This process was facilitated by the United Nations Population Fund Guyana (UNFPA Guyana), the Director of Youth in the Ministry of Culture Youth and Sports (MCYS) Guyana, and Cite Coordinators of Youth Friendly Spaces (YFS)[5] in Guyana. They provided critical logistical support to ensure my access to prospective participants.

Through a purposive/purposeful criterion sampling technique, I selected 24 young Black women from three regions with the highest Afro-Guyanese population. Guest (2006) informs my decision on sample size. He proposes that saturation point is an indicator of sample size for studies without fixed size specifications. However, on analyzing 30 interviews, he reveals that full thematic discovery is possible by the end of the first 12 interviews. Purposive sampling is a non-probability sampling procedure compatible with qualitative research. Purposive sampling is credited for its ability to yield crucially rich information that qualitative-oriented studies require (Patton 2002). 'Criterion sampling' as a strategy of purposive sampling is an approach for selecting prospective participants with predetermined criteria (Patton 2002, 283). As a valid procedural strategy, it eliminates the discovery of criteria problems during the process of an interview, minimizes time, and produces a highly homogeneous sample. The criteria for this study were that the women self-identified as Black, were unmarried, non-cohabitating, sexually active, between the ages of 17 and 25, and sexually involved with men with the expectation of material benefits.

Feminist methodological and epistemological approaches are described as systems of ideas or structures that explain, justify, and guide feminist actions (Jaggar 1984) that orient this work. As such they allow for the integration of research methods (Oakley 1998) while urging researchers to be conscious of the relations between knowledge and power, advises that useful feminist research methods are sensitive to power relations and based on their appropriateness for one's research. Considering the ways in which gender influences what we have come to claim as 'truth,' feminists have answered the question of who can be 'knowers' by producing alternative perspectives that legitimize women as 'knowers' (Harding 1987; Cook 1986; Doucet 2006). Harding urges women to begin their research from the lives of women and broadly with marginalized groups regardless of the researcher's location (1987). Situating Black women's experiences at the center of analysis is important for Black feminist theorizing and my obligation as a Black Caribbean woman to critique the discourse positioning Black women's sexual bodies in particular performative spheres.

Utilizing intersectionality as a framework of analysis, Black feminists have identified the experiences of oppression of racially subordinated women/groups (Collins 2000; Crenshaw 1991; Hulko 2009). As a fundamental position, Black feminists argue against an analysis of identity categories such as race, class, and gender separately or within traditional frameworks. To do so is to obscure the combined effects of discriminatory practices in Black women's lives (Crenshaw 1991). Employing intersectionality as a lens helps this work to explicate young women's perception of and experience with race, gender, and sexuality and how young women's race and sexuality are framed within relations of power.

Race and Gender: A Conscious Embodiment

The experiences of intimacy and economics reproduced in this work derive from 24 self-identified black women of varying social classes. Two of the women are of mixed African and Indian ancestry. Six of them are 19 years old, and all but one live with parents or relatives. They attained secondary education at varying levels, seven are enrolled in tertiary-level education, six are employed, and 12 of them are mothers.

Blackness as a racialized identity is socially and politically interconnected to the Black women's past and present. And though the negative aspersions on their Black physicality remain a point of struggle, they understood themselves as black and woman through the love of the Black self, and pride in their Black feminine aesthetics. For example, Anna responds to the question, how do you feel being a Black woman? She declares:

Anna: For me is just like a beauty queen or princess or something like that.
Int: What about your Blackness makes you feel like a queen?
Anna: Since I was growing up, my mother told me that Black is beautiful, so when I look in the mirror, I know that I am beautiful. It is normal to think so of yourself, so you see when I walk on the road, people respect me.

Rose elaborates:

Rose: To me I aint racist or anything but I feel good as a Black person.
Int: What does feel good mean?
Rose: Well, I am proud to be Black.
Int: What are you especially proud of?
Rose: I like ma' skin color and we shape and I like how Black people does dress.

Int: How do they dress?
Rose: You don't see, we does dress better than the other races. They can't stop we wid dat. I got to be proud because dem other people ain't shape like we, and don't look like we when we dress.
Int: Shape like how?
Rose: You funny, leh me turn round and show you, look!

Ana and Rose articulate their love of the Black self from different points of awareness. Ana cultivates a sense of Blackness as intrinsic worth, while Rose promotes an aesthetic of blackness, from which she perceives her value in a Guyanese cultural milieu. Representing her blackness with the identity of 'queen' Anna rejects a Blackness positioned as 'Other.' Self-affirming as queen transcends her Black embodiment and internally validates her personhood and members of her race. Rose punctuates her love of Blackness with a disclaimer of racism as though it were suspect to love her Blackness without seeming racist. Her love of her Black identity derives from her sense of a unique Black physical body. Her skin tone and her shape are points of Black beauty, which in turn makes her both valuable and superior. The privilege she bestows on Afro-Guyanese women for stylizing the self, within a race comparison, produces the Black female body as both an identity and a performance. However, though she disclaims any inclination of racist sentiments, she seemingly projects an ethnic bias in believing that Afro-Guyanese women's manner of dress outshines the other races in Guyana. In doing so, she ignores the variability of style and its association with culture, class, race, individual preference, and financial means.

Iris however offers an alternative response to the question of her Blackness.

Int: How do you feel about yourself as a Black woman?
Iris: No problem, but I feel bad as a woman and not even the Black because when you're alone, ya can't get wuk and can't provide for ya children it is bad, and right now is bad.

Iris, as a consequence of her experience with poverty, situates her Blackness as peripheral to her gender, and her identity as a mother and sole provider for her children. Poverty, which defines the limits of her ability for childcare, has important implications for her Black single maternal identity. Iris's experience as a single mother and provider for her child may be similar for other women where motherhood is experienced in a context of class and cultural relations. However, as a Black Caribbean mother, the categories single and poor are unlikely to relieve the cultural expectations upholding a maternal obligation of self-sacrifice. This expectation creates a troubling paradox for Afro-Caribbean mothers. While Black mothers are expected to self-sacrifice, they are often under-valued as women, since the bodies of Black single mothers are likely to be read as sexually deviant.

Consciously embodying one's Blackness in self-affirming ways does little to ease the perpetuation of colonial narratives embedded in Guyanese society. Pam shares her experience of race in Guyanese society.

Int: Tell me, how do you think your society sees you as a young Black woman or Black women generally?
Pam: Society see Black people as lower class.
Int: What do you mean by lower class
Pam: Like they just 'deh down'[6] you know when bad things happen the first thing people does seh, that is not a coolie[7] man, it must be a Black man. And when good things happen like CXC results they say it got to be a coolie man child, no Black man can't do that.
Int: Who are the people who talk about Black people 'deh down'?
Pam: De coolie people, but some of de Black people does say de same thing.
Int: How do you feel that people think of you like that?
Pam: Me, it doesn't move me.

Likewise, May explained,

May: The people in this country see Black people as warrish and a mediocre race. They see us as a people with poor business skills. They would say that we can't run business. There is a stigma to being Black.
Int: Why do you think so?
May: Because of the bad way people talk about Black people and the good way they talk about Indian people.
Int: How do you feel about that?
May: Personally, I am not interested in what they say because I have high esteem and their opinion don't matter to me.

The perception that Indo-Guyanese are the standard citizens to which Afro-Guyanese are compared emanates from the colonial assignment on non-white bodies. An ascription of a higher-order social rank to one race forges a struggle for legitimacy occasioned by the discord planted by the British Colonial powers. Williams (2001) and Puri (1997) point to the resonance of racialized symbols including that of the wandering idle African and the hard-working industrious Indian. Puri perceives those racialized tropes as a conduit for hostility between Afro and Indo-Caribbean people (1997).

Yet, the peculiarities of race and class relations in Guyana may be further examined through class formations imposed both by colonial and post-colonial state systems (Danns 2014). Andaiye similarly accounts for those formations following emancipation and post-independence. Black and colored people assimilated the class-related symbolisms and values of the British (Scott 2004). Black people's control of the public sectors allowed for their ascendence to ruling class status and top-level employment in public service, and state economic enterprise (Danns 2014). The class-race dynamics shifted at the point at which East Indians became doctors and lawyers. Their involvement in private enterprise allowed for upward social mobility (Smith 1984; Scott 2004). Guyanese centrality to leading political parties also shifts the race, class, and power dynamics (2004). While those with political power tend to economically improve their race-identified group, the professions, education, socially approved acquisition of wealth, or family background are the main indicators of middle-class Caribbean status.

Afro-Caribbean Women's Sexuality and the Politics of the Body

Non-consensual or consensual sex generally occurs within relations of power and agency which shape gendered practices. Among the several classifications of sex, the women define sex primarily as the heteronormative act of penile-vaginal intercourse. Though this definition is limiting, sex is operationalized in this work as conceived by the young women. The implications of sex produced anxieties for the young women navigating the limits of consent, dissent, responsibility, danger, and desire. These anxieties include the fear of pregnancy, the need to conceal their sexual activities from family members, and the worry about sexual exploitation. Thus, those with the inclination for sex had to imagine the 'appropriate' set of actions since anecdotal accounts of Afro-Caribbean women reveal their cultural conditioning to the social and political value of sexuality.

The women engage sexually with men they label 'boyfriends' and they are 'girlfriends.' The title 'boyfriend' is bestowed after evidence of qualifying characteristics. For example, mutual attraction and symbols of material resources are important prerequisites for the potential boyfriend. However, women's willingness to have sex depends on any combination of mutual affection, symbols of money, attachment, faithfulness/monogamy, trust, and communication. Though the boyfriend's fidelity is not a high-order requirement, his ability to access material resources is a primary condition for sexual consent. The following two extracts represent a typical theme:

Int: Tell me, what are some of the things that made you decide to have sex with a man?
Kim: Well, I did like he, and know that he like me and he was working.
Int: Is working important?
Kim: Like when we go out or like how he would carry me out, and not buy anything for me!
Int: You saying that he should give or buy you things when you go out?
Kim: Well, if I have to give everything wha' I got he around fo?

Kandi responded in this way:

Int: What are the things that made you decide to have sex with a man?
Kandi: I got to like he and he had to run behind me.[8] He got to show me big time interest.

Int: Why do you want him to run behind you?
Kandi: If a man really likes you, he would treat you nice, and don't hurry to have sex.
Int: What is treating you nice?
Kandi: Is when I don't have to ask him or wait long for him to bring something for me. He would give me, and not hold out. And when I tell him me and my friend or my sister or anybody going out, he would ask if I want something.

Kim and Kandi's resistance against a relationship deprived of economic resources is reminiscent of a once popular Caribbean calypso, '*No Money No Love*' (Sparrow 1969). In the song, a couple separated because of the man's ill fortune. The woman in the relationship questions her future with a man without financial resources, while she sings the refrain, 'You can't love without money.' The notion of divorcing intimate relations from economic considerations in order to flourish fails to consider arranged marriages in parts of the world and in Caribbean societies. Such arrangements function as economic investments to ensure the prosperity of families. Further, contemporary consumption culture requires economic considerations and activities of leisure, as in the case of Kandi and Kim 'going out' to improve intimate relations.

Yet, the assumptions accompanying 'nice' treatment open a dialogue on how culture may potentially facilitate women's exploitation of men's financial resources. Admittedly, economic support from men allows single women to narrow the gaps in social inequalities, as evident in Kim's question about the boyfriend's role if she gives everything. Given that the intertwining of sex and economics are products of cultural relations, then ideally a culture that socially and politically valorizes sexuality will teach women to exploit its value for themselves.

Young women's intimate relations are also attached to other kinds of expectations, Joan explains:

Joan: I was friends with a boy who said he like me. He ask me for sex and I din want, but my friend said that he is a nice boy, and if don't have sex with him he might find another girl, so I decide. And the next day he told all his friends. I saw him and curse him out.
Int: Why did you curse him out?
Joan: I was shame because I give him my virginity and he tell everybody and I never talk to him again.
Int: Ok, but tell me what does 'nice boy' mean?
Joan: Is a boy who got respect for girls and don't go running he mouth to he friends, like a auntie man.[9] Is only woman who does talk so. Nice boys got respect and like, if you ask him to do anything for you, he would.

Joan is one of the few young women who engage in sex on account of peer pressure. She neglects an active practice of a politics to engage in a sexual agency that honors her feelings about her sexual desire. And so, her negligence results in disappointment and shame. Regrettably, gender normative behaviors that teach young women to deal with sexual 'transgressions,' through shame or in secret is burdensome. Sexually violated or exploited women are likely to hide their abuse from a society that discounts their oppressive experiences. While gender roles are fluid, scripting the nice boy as a protector of feminine virtue and support to women is partly explained within romanticized notions of the gentleman. Young men who have internalized a dominant masculine gender identity are likely to perceive a sexual encounter as currency for peer validation, and as male conquest that maintains a hierarchy of masculine dominance. Conversely, Joan's feminization of the nice boy for his failure to operate within her gendered frame of reference likely derives from her culture's construction of an 'appropriate' masculine identity. Masculinity, defined in relation to femininity, for example, as 'talkative' threatens a manly ideal, thus non-confirming men become targets of prejudice. Further, the view of women as talkative reflects a deep-seated myth that supports the devaluation of women's voices. Paradoxically, men's voices when used excessively, extensively, or repeatedly in public and private spaces are deemed as powerful.

Sex is a (dis)embodied practice of power for women in the study, especially since they identified men's sexual desire for them as an important indicator of men's love. The women generally agreed that their sexual satisfaction comes from the attainment of orgasm. This orgasm was explained through the metaphors 'when I come,' 'when I break,' and 'when I feel nice.' The women shared:

Int: Do you gain sexual satisfaction in your present relationship?
Joan: I am not sure if I get orgasm, so when he ask, I just say I like it and I fake the orgasm so that he could 'come.' Anyway, the most important thing I like in our relationship is respect. He respects my family and me.

Jackie responds to the same question:

Jackie: I don't pay attention to that anymore, I 'come' sometimes. When it don't happen, I don't always tell him because sometimes he listens and sometimes, he leaves me like that and he is finished and I don't tell him, I let him finish.

However, this is not true for all women Rea reveals:

Rea: I used to be frighten to say anything because he was my third boyfriend and I never used to come with him.
Int: What were you frightened about?
Rea: I din want to say I don't feel nice because he is lil ignorant and would tell me that I must be had a lot of men. So, I wait and this thing boiling up in me,[10] till I can't tek it no more, so one day in the middle of it I just scream out and come out from under he.
Int: And what happened?

Rea: I just start to quarrel and tell he everything. He went home and I din see he for four days, then he come back and say ok we gon do it different. So now he does wait til I tell him I ready and then he does enter.

By these accounts, sexual satisfaction is challenging for women, who privilege men's sexual needs, over their sexual pleasure. The fact that some women fail to negotiate their erotic pleasure while other women engage in sexual agency suggests that the cultural privileging of Caribbean men's sexuality does not solely account for women's self-denial. Women are likely to assimilate patriarchal values and its active/passive hierarchy of sexual roles to validate the boyfriend's masculinity and inform their sexual silence. Joan's accommodation of the boyfriend's respect instead of her satisfaction supports the position that many women in heterosexual intimate relations resolve their lack of sexual satisfaction by transferring that longing to other aspects of their lives (Barriteau 2011). At the same time, the reliance on the penis for sexual satisfaction enacts sexuality through a masculine lens, thereby reinforcing men's privilege to define what sex is or how it should be. Reliance on the penis also gives to men the primary responsibility for women's orgasms rather than women themselves. The fact that nearly half of the women admit to faking orgasms contradicts their collective agreement that their sexual satisfaction is important.

Noteworthy however is Rea's active quest for sexual satisfaction. It captures the dynamism of sexual agency and the strategies enabling women's autonomy in sexually oppressive relations. Her actions expose the performative possibilities of women's bodies when they resist the instrumentality of their sexuality to self-define as fully conscious sexual subjects. Women who claim the space for sexual agency, make sexual pleasure possible and contribute to the de-normalization of repressive practices that men have come to accept as valid.

Passion and Possibilities: Not All about Love

The theorizing of romantic love from multiple perspectives allows for competing meanings, among different cultures. While the women reveal that they are 'in love' with or 'love' their boyfriends, 'being in love,' or 'love' reflects similar sentiments. Within various levels of emotional intensity. The women defined their love as: 'to like a person real bad,' 'to always want the person around because they make me feel good,' 'the feeling I get in my stomach and belly when I see my boyfriend,' 'liking somebody and would do anything for him,' 'feeling good with somebody when you have sex and the person is not selfish and have you alone,' 'doing things to show special interest in him,' 'staying with the person, even though they does really fret you,' and 'when I see my boyfriend talking to a girl and I get jealous.'

By the women's account, their feelings of love for the boyfriend include a strong attachment, unconditional support, mutuality in sexual relations, faithfulness/monogamy, and consideration. Conversely, they interpreted the boyfriend's love through his sharing of material resources, sexual desire for them, and care. Jane and other women, however, admit that their boyfriends' provision of money is based on other considerations. Jane states:

> I think that he wan' me to think that no other man won't give me so much money, because he would count the money in front of me when he give me. He would say, I am giving you fifty thousand dollars or how much ever. He always put the money in my hands then say.

Many women are conscious of the instrument and affective dimensions of men's money/gifts; men want something and men care. Caribbean men in non-commercial sex relations with women tend to cultivate affective relations and sustain women's interest by giving them money (Sobo 1993). However, the consideration of care did not erase women's perception that money strategically promotes men's interests to create or maintain relations of power.

For these women, the point at which men offer money or gifts holds importance for the development of trust and a sense of 'belonging.' Generally, men offer money or gifts early in their acquaintance with the women; however, the women having to ask for money proved a point of contention. Lorna expresses her feelings about making a request for money.

Lorna: If I have to ask, I won't feel too nice because I would think that he not acting like a man, he mean[11] or something and he not looking out for me.

While. Fiona strategizes:

Int: What has your boyfriend given to you?
Fiona: Money and gifts.
Int: Did you expect to receive these things?
Fiona: Yes, especially when he start to work.
Int: Who decides what you will receive?
Fiona: Usually he would ask, wha ya got to get?
Int: Do you ask?
Fiona: No, and he don't give every time, but sometimes I would throw a hint, I would tell my little brother to come and ask me for money, so he could hear I aint got.

For Lorna having to ask for money denotes a weak masculine performance and a lack of care and Fiona reconciles her discomfort to request money with the duplicitous involvement of her brother. Evident here are traditional gendered beliefs shaping women's notion of men's role. In these relations a direct request to men for money appears to devalue the women and undermines the masculine. Though Afro-Caribbean women participate in the labor force at varying levels, an understanding of the male as 'provider' persists for many women.

Beth however is conflicted about the point at which money should be received. Beth states:

> I prefer to get things before we get serious, but anyway, sometimes men give you things quick to bribe you, but then if they don't give before they won't give after. And if you take the things too quickly and don't make styles,[12] they will think that you does always take things quick from men.

Beth has an uneasy relationship with intimacy and economics. She is cautious about engaging in sex without men's demonstration of care, thus her preference for an early gift. At the same time, Beth draws on notions of 'respectability' as she contemplates the unfavorable image that may accompany the woman who readily accepts gifts from men. Beth is clear about the potential of an early gift to shape women's thoughts and feelings; it functions as an inducement to influence women's sexual choices. The anxiety experienced by many young women over men's eager offer of material support to initiate relationships, speaks to the contradictory ways in which men and women 'gender' intimacy and material transfers in sexual relations.

The perception of men's ulterior motive for contributing money does not affect the centrality of the boyfriend's money, to 'love' and to sexual access. Yet these women claim unconditional support to men as a characteristic of their love. A question on the meanings they assign to men's denial of a gift or money becomes important. Janet and Jill clarify the political significance of such a denial.

Int: What happens if he stops giving money?
Janet: I would ease out, and send back the clothes he does ask me to wash because I would think that he just want to use me and I would pull myself away.[13]

And Jill recalls:

> A boy do that to me already. He say he cant do better so I say I will hold on for a lil time, but a night he went in our toilet, and a lot of money drop out he pocket and my lil sister find it. I get so mad because when he come, he always feeling me up,[14] and always want something to eat. My mother does say give him, and he got money and he not giving, he wan use me. I tell him don't come back because he ain' using me.

Janet's willingness to return unwashed clothing and Jill's termination of alimentary support points to the value that many single young women attach to their care labor. Men's denial or withholding of money undermines single women's relational expectations, positions those relations at the margins of reciprocal care, and makes problematic the women's unconditional support. In fact, as Zelizer observes when cultural pretensions are stripped from sexual relations, it exposes people's need for returns on their material and emotional investments (2005). Men's withdrawal of money while expecting single women's care labor likely derives from sexist notions that care labor resides in the domain of the feminine. More important, the political significance of undoing domestic support speaks to young women's vigilance against exploitative relations. Anecdotally, Afro-Caribbean women perceived to be 'used' by men are ascribed an identity by peers that affects how they navigate their lives in their communities. They are labeled as stupid, of low self-value, careless with the self, shameless and their bodies become a site of public slander.

Speaking with one voice, the women under study would not forego money from their boyfriends even with individual financial security. They reasoned that to do so relieves men of their obligation, even as men maintain expectations of sex. It is not unreasonable to assume that the meanings they attach to men's money converge around beliefs of 'entitlement' and 'belonging.' Affirming their social location as 'girlfriends' infer a position of privilege that accompanies a sense of financial entitlement. The status of 'girlfriend' legitimizes the relationship and creates the social conditions and characteristics allowing for their sense of entitlement.

The distinction between transactional sex as 'an exchange of sex for money or goods' and women's understanding of money and goods provided in relations that are sexual was clarified through two hypothetical questions. (a) Imagine the person you have sex with visits you repeatedly, has sex goes home soon after but leaves money for you, how would you respond? (b) How would you respond if he visits, demonstrates affection, and gives money but does not have sex? The women in this study strongly reject the idea of sexual relation that replicates a direct payment for sex. They agreed that sex and money without affective attachment borders on disrespect and belong to the domain of the prostitute. And such an experience would suggest the boyfriend's loss of respect. Further, they argued that affective expressions must involve the commingling of sex and economics in their 'boyfriend/girlfriend' relations. Their response to the second question was articulated around feelings of rejection, vulnerability, betrayal, sadness, and an assumption of the boyfriend's reluctance to commit. The women claim that sex with the boyfriend signifies relations of love, fidelity, attachment, and sexual desirability. Thus, its withdrawal would define their relations as platonic or benevolent and the loss of their potential to exercise power or resistance over the boyfriend. The constructs allowing for the meanings women impute to the boyfriend's potential retreat from sex imply that they consider sex as a relational practice, that is, sex occurs in the context of a relationship.

Sex and Transactions Ruptures and Connections

Few people might disagree that Afro-Caribbean men providing money to single women are not necessarily performing from a position of benevolence. Yet we may not know enough about the deeper nuances of young Afro-Caribbean women's non-conjugal relations of intimacy and economics. To provide context to their relations, I briefly examine the characteristics of prostitution as one of the relations in which sex and money are exchanged.

Caribbean studies of non-conjugal sexual relations have determined their forms within visiting unions (Danns 1989), sex work (Kempadoo 2004), or have placed those relations on a spectrum from love to prostitution (Chevannes 2001). In the Anglophone Caribbean non-conjugal, non-commercialized relations in which sex and money converge often rely on intimacy to create emotional ties and fidelity to be maintained by women.

Prostitution as an act that commingles sex and money entails an offer and acceptance of material rewards for sexually gratifying experiences. Thus, the money or goods paid to those who sell sex purchase a commodity. Magnanti (2012) illustrates the contours of prostitution as a bell curve with a desperate hard end, a big middle, and an even more privileged end. The contours suggest a diversity of services with disparities in income and skills, to ensure that commercialism reigns over sentiment. Sex work in the Anglophone Caribbean is legally prohibited (Robinson 2008). It is seen as disruptive to the social order of 'appropriate' sexual practices. Contextualizing prostitution within socially disruptive acts likely informs the position that 'prostitutes refuse to mystify the instrumental dimensions of sex' (Sobo 1993, 184). Prostitution codified as foreclosing the expressive dimension of sex renders it a mere act of sex and money exchange.

The sexual relations of the women who participated in this study are not prohibited through legal sanctions, nor negotiations of an offer and acceptance of sex for money. Rather than detachment from intimacy to maintain rigid divisions between their public and private lives, a series of culturally tolerated procedures cements affective ties, care labor, and self-imposed compliance to monogamy. Though money, as a resource is integral to sex, their affective ties mediate their bodily experiences from the social and sexual tensions likely to occur with women who exchange sex for money. As such their sexually embodied practices cannot be explained within the prevailing definitions of transactional sex.

Magnanti's illustration of prostitution or transactional sex within stratified statuses and hierarchical privileges marks an important contrast with women's practices. Chavannes's 'negotiated sexuality' which he positions between 'love' and 'prostitution' in intimate and economic relations of young Caribbean people (2001:199) may share features with 'tacit sexualities.' Though the various relational categories on Chavannes's continuum appear to involve a different set of core relational practices. Common to 'tacit sexualities' and 'negotiated sexuality' are practices that support the needs of men and women within reciprocal socio-sexual interests, and in many instances based on 'mutual attraction,' 'romantic love,' and 'gender-based commitment' (196–199).

Of the relations where intimacy and economics converge, my work locates the women's relations within the various forms of 'romantic' non-residential relations. These relations are neither purchased nor outside the scope of material considerations, rather they are characterized by interpersonal interactions combined with material objects like 'food, books, and jewelry' (Zelder 2009, 135). The combined characteristics of the intimate and economic practices of women in this study I categorize as 'tacit sexualities.'

Transformative Narratives: From Transactional Sex to 'Tacit Sexualities'

'Tacit sexualities' as defined earlier is an alternative conceptual framework that challenges the problematic use of the term 'transactional sex' on young Caribbean women's sexual bodies. 'Tacit sexualities' describes how many Afro-Caribbean women navigate their intimate and economic relations. Situated within relations that commingle intimate and economic engagements, 'tacit sexualities' is socio-sexually positioned outside of conjugal relations or those legally and culturally designated as prostitution.

The concept 'tacit' describes the sexual politics within which single Afro-Caribbean women navigate the economic and sexual aspects of their intimate relations with men. The expectation of material transfers from men to women in these relations is implicit. 'Tacit' as the women's mode of managing their relations is 'silent' but not 'secret.' The 'silence' in 'tacit sexualities' constitutes an 'unspoken' 'knowing,' that is recognized and interpreted by two sexually interested heterosexual people that sex and money must coexist in intimate affective relations. And that a set of mutually interpreted expectations follow or precede intimate sexual relations. In this manner the 'knowing' is not secret; it is productive.

This 'knowing' that mediates the material transfers of Caribbean men is partly shaped by conventional gender roles, manifested as a demonstration of men's interest and functions to enable the attention of women. The characteristics of *'silent'* but not *'secret,'* informs a politics on material transfers which in turn enforces the boundaries within which the intimate and economic relations are defined, understood, and experienced by the women. A function of the boundaries acts to deter financial negotiations attached to sex. The boundaries also buffer against potential exploitation since men's transfer of material resources is essentially arbitrary. The relations are solely regulated by the individuals operating within a set of cultural influences and personal beliefs.

'Tacit sexualities' is modeled around three structuring pillars, 'Sex,' 'Money,' and 'Love.' 'Tacit sexualities' as an amalgam of resources and relations are nuanced, contextual, and multi-layered with connecting and overlapping meanings. Each structuring pillar holds specific and intersecting meanings. 'Love' as a structuring pillar is infused with a set of feelings and behaviors, for example, women demonstrate their 'love' through sacrifice, surrender, nurturance, and domestic support, and experience it through a feeling of safety or contentment. Yet 'Love' intricately interacts with sex and money and provides a context for normalizing these relations.

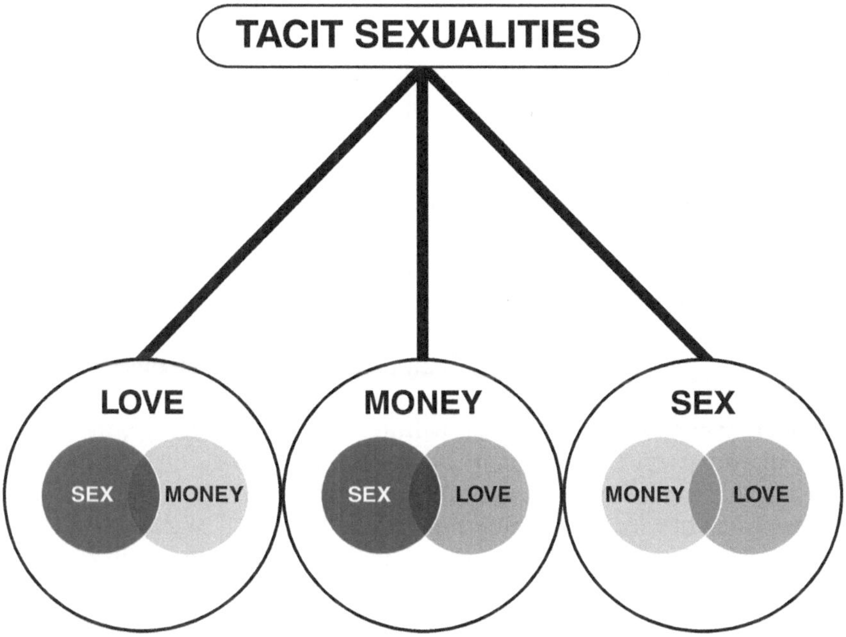

Figure 10.1 'Tacit sexualities'.

'Sex' as another relational pillar is experienced through penile-vaginal connection, yet sex gives meaning to money and love. Similarly, 'Money' as a critical resource holds important implications for sex and notions of love.

A critical attribute of 'tacit sexualities' as a lens for theorizing on young Afro-Caribbean women's sexuality is its principle of equal value on which the pillars function. Namely, the relations of 'Love' or 'Sex' neither minimize nor increase the status of 'Money' as a resource. The boundaries of Sex, Money, and Love are interconnected in a manner that the withdrawal of one core pillar disrupts or demolishes the relations and by extension the complete framework under which their relations are constructed and experienced. Accordingly, if 'Love' is removed from the relations those relations are deemed as prostitution. Likewise, if 'Sex' is removed from the assemblage the relations are considered platonic, benevolent, or altruistic while the withdrawal of 'Money' defines the relations as exploitative or oppressive. Though 'Money' and 'Sex' as two independent pillars can function efficiently in prostitution, 'tacit sexualities' depends on and intersects with each pillar to maintain its relational integrity.

Conclusion

'Tacit sexualities' is one of the diverse forms of relations connecting sex and money in the Anglophone Caribbean. It involves a set of practices with relevance for a particular space and demonstrates how people in different spaces comprehend and navigate intimate and economic relations. 'Tacit sexualities' deviates from the relations defined as 'transactional sex' or prostitution and exposes the complexities of sexual practices as well as the problematics of attributing homogenous meanings to sex and monetary transfers.

Still, the combined meanings of the interdependent pillars maintaining 'tacit sexualities' can obscure the dynamics of gendered power. For example, the attachment of money to sex and its relatedness to notions of love can disguise both its sexually objectifying propensity, and the strategic abuse of power that men's money can serve. To the extent that money is infused with meanings of love and sex is the threat that sex can be unwittingly exchanged for money by those who seek love.

Though new and different sexual cultures continue to emerge, the contemporary discourse on transactional sex maintains its fixity on race and place away from European and North American localities. Yet, the ever-evolving propensity of knowledge can undermine the influence of hegemonic discourse over time. Theorizing that emanates from Black feminist spaces offers multiple possibilities for new conceptions of Black women's sexual practices to be imagined.

Notes

1 I explain 'intimacy' as a heterosexual sexual relationship of significant emotional/romantic importance, where the parties share confidence, and trust, offer comfort, and enjoy a sense of security.
2 Any reference to Black-Caribbean or Afro-Caribbean women broadly refers to Black women in the Anglophone Caribbean.
3 My use of the term young Afro-Guyanese refers to Black people in Guyana.
4 The references to money signify both currency and/or valuable goods. The same meanings are applied to material transfers/benefits.
5 YFS are spaces designated by MCYS to accommodate youths to socialize and for personal development training.
6 The expression 'deh down' in this context suggests low morals, low class, or poor quality of life.
7 Coolie is a commonly used expression in Guyana that describes those of East Indian Ancestry. Though the term is generally used in non-complimentary ways, in informal and rural community spaces the expression is tolerated when used inoffensively.
8 Run behind me means to pursue me.
9 Anti-man, generally pronounced 'auntie-man' is a derogatory Guyanese colloquial expression for the homosexual male or for men who display traits that Guyanese do not consider masculine.
10 This thing boiling up in me, refers to her feelings of increased frustration or anger.
11 'Mean' is a Guyanese colloquial expression for a person considered cheap or stingy.
12 Here don't make styles mean to feign indifference.
13 Here, pulling myself away means to create physical and sexual distance from the boyfriend.
14 Always feeling me up implies that the boyfriend constantly makes sexual overtures.

References

Barriteau, Violet Eudine. 2011. "Theorizing Sexuality and Power in Caribbean Gender Relations." In *Sexuality Gender and Power: Intersectional and Transnational Perspective*, eds. Anna G. Jonasdoittir, Valerie Bryson, and Kathleen B. Jones, 75–91. New York: Routledge.

Barrow, Christine. 2008. "Sexual Identity, HIV and Adolescent Girls in Barbados." *Social and Economic Studies*, 57: 7–26.

Chevannes, Barry. 2001. *Learning to be a Man: Culture Socialization and Gender Identity in five Caribbean Communities*. Kingston: University of The West Indies Press.

Collins, Patricia Hill. 2000. *Black feminist Thought: Knowledge Consciousness and the Politics of Empowerment*. New York: Routledge.

Cook, Judith A, and Mary A. Fonow 1986. "Knowledge and Women's Interests: Issues of Epistemology and Methodology in Feminist Sociological Research." *Sociological Inquiry*, 56: 2–29.

Crenshaw, Kimberle. 1991. "Mapping the Margins: Intersectionality, Identity Politics, and Violence Against Women." *Stanford Law Review*, 43: 1241.

Curtis, Debra. 2009. *Perils and Pleasure: Girls Sexuality in A Caribbean Consumer Culture*. New Jersey: Rutgers University Press.

Danns, George K. 2014. "The Impact of Identity, Ethnicity and Class on Guyana's Strategic Culture." *American International Journal of Contemporary Research*, 4: 65–77.

Danns, George K Danns, and Basmat Shiw Persad. 1989. *Domestic Violence and Marital Relationships in the Caribbean: A Guyana Case Study*. Georgetown: University of Guyana.

Decena Carlos Ulises. 2011. *Tacit Subjects: Belonging and Same-Sex Desire among Dominican Immigrant Men*. Durham NC: Duke University Press.

Doucet, Andrea, and Natasha S. Mauthner 2006. "Feminist Methodologies and Epistemology." In *Handbook of 21st Century Sociology*, eds. C. D. Bryant and D. L. Peck, 36–42. Thousand Oaks: SAGE.

Foucault, Michel. 1972. *The Archaeology of Knowledge and Discourse on Language*. New York: Pantheon Books.

———. 1990. *The History of Sexuality: An Introduction*. New York: Random House Inc.

Guest, Greg, Arwen Bunce, and Laura Johnson. 2006. "How Many Interviews Are Enough? An Experiment with Data Saturation and Variability." *Field Methods*, 18: 59–82.

Harding, Sandra. 1987. 'Is there a Feminist Method?' In *Feminism and Methodology*, ed. Sandra Harding, 49–82. Indiana: Open University Press.

Hooks, Bell. 1981a. *Ain't I A Woman: Back Women and Feminism*. Cambridge: South End Press.

Hope, Ruth. 2007. *Addressing Cross-Generational Sex: A Desk Review of research and Programes*. Washington: Population Reference Bureau.

Hulko, Wendy. 2009. "The Time and Context Contingent Nature of Intersectionality and Interlocking Oppressions." *Journal of Women and Social Work*, 24(1): 44–55.

Hunter, Mark. 2002. "The Materiality of Everyday Sex: Thinking Beyond Prostitution." *African Studies*, 61(1): 99–120.

Illouz, Eva. 2006. "Romantic Love: Interview with Eva Illouz." In *Handbook of the New Sexuality Studies*, eds. Nancy Fischer, Stephen Seidman, and Chet Meeks, 40–48. London: Routledge.

Jaggar, Alison M., and Paula Rothenberg. 1984a. *Feminist Frameworks: Alternative Theoretical Accounts of the Relations between Women and Men*. New York: McGraw-Hill, Inc.

Kaufman, Carol E., and Stavros E. Stavrou 2004. "Bus Fare Please: The Economics of Sex and Gifts Among Young People in Urban South Africa." *Journal of Culture, Health & Sexuality*, 6(5): 377–91

Kempadoo, Kamala. 2004. *Sexing the Caribbean: Gender Race and Sexual Labour*. London: Routledge.

———. 2011. "Power, Labour, Pleasure: Sexuality in Everyday Life." In *Working Paper Series*, 15. Cave Hill University of the West Indies.

Kempadoo, Kamala, and Andy Taitt. 2006. "Gender Sexuality and Implications for HIV/AIDS in the Caribbean: A Review of Literature and Programmes." Bridgetown: UNIFEM and IDRC, 2–55.

Leclerc-Madlala, Suzanne. 2003a. 'Transactional Sex and the Pursuit of Modernity.' *Social Dynamics*, 23: 213–33.

Luke, Nancy, and Kathleen Kurtz. 2002. "Cross-Generational and Transactional Sex relations in Sub Saharan Africa: Prevalence of Behavior and Implications for Negotiating Safer Sexual Practices." Washington: USAID.

Maganja, R. K., S. Maman, A. Groues, and J. K. Mbwambo 2007. "Skinning the Goat and Pulling the Load: Transactional Sex among Youth in Dar es Salaam, Tanzania." *AIDS Care*, 19 (8): 1–27.

Magnanti, Brooke. 2012. "Dr. Brooke Magnanti – Scientist and former prostitute." In *Hard Talk*, ed. Katya Adler. BBC World Services. Accessed October 2015: http://www.bbc.co.uk/iplayer/episode/p00ylyf2/Hardtalk_Dr_Brooke_Magnanti

Noble, Ronald. 2010. *The Globalization of Crime: A Transnational Organized Crime Threat Assessment*. Vienna: UNODC.

Oakley, Ann. 1998a. "Gender Methodology and People's ways of Knowing: Some Problems with Feminism and the Paradigm Debate in Social Sciences." *Sociology*, 32(4): 706–31.

Patton, Michael Quinn. 2002. *Qualitative Research and Evaluation Method*. California: SAGE Publications Inc.

Phillips, Daphne. 2006. "Youth HIV/AIDS in the Caribbean: Teenage Sexuality in Montserrat." *Social and Economic Studies*, 55: 32–64.

Puri, Shalini. 1997. "Race, Rape and Representation: Indo-Caribbean Women and Cultural Nationalism." *Cultural Critique*, 32: 119–63.

Robinson, Jonathan, and Ethan Yeh. 2011a. "Transactional Sex as a Response to Risk in Western Kenya." *American Economic Journal: Applied Economics*, 3: 35–64.

Decena, Carlos Ulises 2008. 'Tacit Subjects', *A Journal of Lesbian and Gay Studies*, 14: 339–59.

hooks, bell. 1981b. *Aint I A Woman: Back Women and Feminism*. South End Press: Cambridge.

Jaggar, Alison M and Paula Rothenberg. 1984b. *Feminist Frameworks: Alternative Theoretical Accounts of the Relations between Women and Men*. New York: McGraw-Hill, Inc.

Leclerc-Madlala, Suzanne. 2003b. 'Transactional Sex and the Pursuit of Modernity', *Social Dynamics*, 23: 213–33.

Oakley, Ann. 1998b. 'Gender Methodology and People's ways of Knowing: some Problems with Feminism and the Paradigm Debate is social Sciences', *Sociology*, 32.

Patricia, Hill-Collins. 2000. *Black feminist Thought: Knowledge Consciousness and the Politics of Empowerment* (Routledge: New York).

Robinson, Jonathan, and Ethan Yeh. 2011b. 'Transactional Sex as a Response to Risk in Western Kenya', *American Economic Journal: Applied Economics*, 3: 35–64.

Swidler, Ann, and Susan Cotts Watkins. 2007. 'Ties of Dependence: AIDS and Transactional Sex in Rural Malawi', *Studies in Family Planning*, 38: 147–62.

Rowley, Michelle. 2000. "Reconceptualising Voice: The Role of Matrifocality in Shaping Theories and Caribbean Voices." In *Gendered Realities: Essays in Caribbean Feminist Thought*, ed. Patricia Mohammed, 22–43. Kingston: University of the West Indies Press.

Scott, David. 2004. "Counting Women's Caring Work: An Interview with Andaiye." *Small Axe*, 13: 123–217.

Smith, Raymond T. 1984. *Culture, Race and Class in the Commonwealth Caribbean.* Kingston: Department of Extra-Mural Studies.

Sobo, Elisa. 1993. *One Blood: The Jamaican Body*. Albany: State University of New York Press.

Sparrow, Mighty. 1969. "No Money No Love." In *Mighty Sparrow Volume 1. Freak Lyrics*. London: Sanctuary Records Group.

Stark, Laura. 2013. "Transactional Sex and Mobile Phones in a Tanzanian Slum." *Journal of the Finnish Anthropological Society*, 38(1): 12–36.

Stoebenau, Kristen. 2011. "More Than Just Talk: The Framing of Transactional Sex and its Implications for Vulnerability to HIV in Lesotho, Madagascar, and South Africa." *Globalization and Health*, 7(34): 1–15.

Swader, Christopher S., et al. 2013. "Love as a Fictitious Commodity: Gift for Sex Barters as Contractual Carriers of Intimacy." *Sexuality and Culture*, 17: 598–616.

Swidler, Ann, and Watkins Susan Cotts. 2007. "Ties of Dependence: AIDS and Transactional Sex in Rural Malawi." *Studies in Family Planning*, 38: 147–62.

UNAIDS. 2009. "UNAIDS Guidance Note on HIV and Sex Work." In. Switzerland: Joint United Nations Programme on HIV/AIDS (UNAIDS) 2009–2012.

Wamoyi, Joyce, Angela Fenwick, Mark Urassa, Basia Zaba, and William Stones. 2010. "'Women's Bodies are Shops': Beliefs About Transactional Sex and Implications for Understanding Gender Power and HIV Prevention in Tanzania." *Archives of Sexual Behavior*, 40: 5–15.

Williams, Brackette. 2001. "Ideology and the Formation of Anglo-European Hegemony and Cultural Domination in Guyana." In *Caribbean Sociology: Introductory Readings*, eds. Christine Barrow and Rhoda Reddock, 270–86. Kingston: Ian Randle.

Zelder, Martin. 2009. "The Essential Economics of Love." *Teoria*, 29(2): 133–50.

Zelizer, Viviana A. 2005. *The Purchase of Intimacy*. New Jersey: Princeton University Press: New Jersey.

11 University Plantation Il/logics

Black Women's Fugitivity and Futurity in the Wake of COVID-19 and the Global Anti-Racist Uprisings of 2020/21

Andrea N. Baldwin

Introduction

In her 2013 article "Plantation Futures" in *Small Axe*, Katherine McKittrick argues that the plantation, while casted as a "backward" institution long remaining in the past, actually "moves through time, a cloaked anachronism, that calls forth prison, the city," (9) and, as I argue in this chapter, the university.[1] By moving through time, McKittrick invokes the plantation as a palimpsest upon which our contemporary institutions of power, policing, and surveillance are written. The markings, the spirit, and the geography of the plantation remain. However, as sure as this colonial legacy is embedded within our society today, so too remain the methods and strategies developed and used by those who have had to survive the plantation's historical brutality. This, what McKittrick refers to as plantation logics, does three things: identifies the ways in which Black oppression, exploitation and death have become normalized due to the workings of the plantation; demonstrates how we are collectively complicit in reproducing this system as normal and natural; and makes space to imagine and create alternative systems of knowledge, not purely oppositional to that of the plantation but one that envisions "a future where a correlated human specific perspective is honored" (Mckittrick 2013, 11).

In this chapter, I utilize McKittrick's plantation logics to map how the colonial underpinnings of the university have traveled through time such that Black people have always had to function as fugitives within the space. Herein, I examine the ways in which Black folx within the academy have and continue to enact "the quotidian practice of refusal" (Campt 2014), the refusal to be refused (Moten 2018), the refusal to settle (Halberstam 2013), and the possibility to live unbounded lives Gilmore (2022). I am particularly interested in Black women and how they have historically been able to (dis)locate Black survival outside of the spatiotemporality of the institution through, among other things, placemaking, knowledge production, translation and sharing, network building, archiving and historicizing, and movement building toward futurity. Black feminist Tina Campt theorizes futurity as a striving for possibilities of the future real conditional, or that which would have had to happen for the

DOI: 10.4324/9781003143550-16

future to be realized. According to Campt (2014), this striving implicates us and our present actions in realizing a future such that Black women have always had to imagine and enact a not yet present future in the academy from which we now benefit and can build upon. I posit therefore that engagement with plantation logics as a Black transnational feminist researcher within the academy provides me with ways of rethinking the colonial project that is the university toward a commitment to enacting a decolonial knowledge project which envisions an alternative to the current academic knowledge production system. This endeavor is one that seeks out futurity by engaging in a fugitive project. By that I mean a project which seeks to escape the restricting borders of the academy by imaging knowledge production through the lens of plantation futures as an alternative knowledge destination that shows up in not only Black feminist intellectual work, but also in Black feminist social, material, and artistic productions.

In this chapter, I advance that as plantation logics move through time, they hold within them possibilities of futurity. Using the example of the contemporary Black feminist organization Black Women Radicals (BWR) in this chapter, I also demonstrate how in our lifetimes Black feminists are able to explore these possibilities as we work both outside, within, and across academic space and time to make a way for a more just future. In doing so, I recognize that according to McKittrick, the technologies of oppression, even though in some ways very different from those experienced by Black women of the past, still exists and that persistent plantation logics became even more pronounced in all spheres of Black life, including in the academy, during the COVID-19 pandemic and anti-racist uprising sparked in 2020 by the murders of George Floyd, Breonna Taylor, and many other Black folx. I continue by using Black feminist historical examples of fugitivity to demonstrate how Black women paved the way for us today to inhabit these contemporary fugitive times and spaces. As I draw from these examples of creative theoretical, ideological, ontological, and practical fugitivity to demonstrate how we can move within and beyond the academy's plantation technologies of oppression in its present iteration, I look, like Black feminist fugitives of the past, to how our decisions to move or to stay in place have implications for Black feminist futures. It is in this way that I argue engaging plantation logics are necessary if Black folx are to honor and envision academic freedoms.

The Longue Durée of the Plantation and Its Il/logics

In her 2016 text *In the Wake: On Blackness and Being*, Black feminist Christina Sharpe writes about "that past, not yet past, in the present" (13). In her methodology of redaction and annotation, Sharpe makes it clear to us, as she writes about the ways in which current crises including violence, surveillance, and human displacement are part of the longue durée of a system of colonization which has never seen the humanity of Black people, that we were not meant to survive. Sharpe for example tells of her encounter with the photograph of the little Haitian girl who after the 2010 earthquake in Haiti was made ready for evacuation by someone taping the word ship across her forehead. What Sharpe makes clear through this example is that the history of enslavement, rebellion, punishment, oppression and dehumanization of Black people precipitated the inevitability of this child's labeling. In her orthographic application of language to Black life, Sharpe writes about the hold, and the ways in which the hold of the slave ship has over time become the hold of the state, where Black people being held in the belly of the state, in its prisons, detention centers, schools are kept through "the language of violence: the language of thirst and hunger and sore and heat, the language of the gun and the gun butt, the foot and the fist, the knife and the throwing overboard" (2016, 70).

Like McKittrick, Sharpe's attention to this violence describes the ways in which the logics of the plantation live on in our contemporary period such that for many Black people, to be Black means "fitting the description of the non-being, the being out of place, the noncitizen always available to and for death" (2016, 86). This connectivity between past and present, between the conditions of violence and vulnerability faced by generations of enslaved Africans and their descendants is felt deep underneath the skin in this 21st century of advanced neoliberal global capital, evident in the murders of countless Black people including George Floyd and Breonna Taylor. Black people continue to experience enhanced vulnerability to ongoing and pervasive surveillance, violence, harm, and death within all spheres of society.

In writing about the longue durée of oppression, I am reminded of Sharpe's reference to residence time (2016, 19), particularly as I think about how global capitalism through its continued use of technologies of violence and dehumanization invokes a sense of "perpetual urgency" (Cooper 2016, n.p.) which accelerates Black death through as McKittrick writes, the "normalizing mechanics of the plantation" (2016, 11). This perpetual urgency is seen in the push to return to "normal" as Black people in particular continue to bear the brunt of a pandemic that has ravished the lives and livelihoods of the most vulnerable within our communities. In this time of pandemic and protest, what to Black people is a sense of normalcy as dictated by the pace of whiteness but death? Because our labor has always been essential to global capital, we have always been essential workers, enslaved, unpaid, underpaid, without homes, without space, and evicted.

For those of us who sit in the academy, this push to return to normal, a normal that has for the last few decades increasingly incorporated Black bodies as contingent laborers, and which perpetuates our vulnerability through adjunctification, puts our lives at great risk. This is despite changes in the ways in which the academy has re/organized, re/structured, re/branded itself to appear more inclusive of Black people and Black women, and the ways in which it em/de/ploys technologies and techniques which it espouses are equitable and seeks to redress past violence committed by the academy. Carole Boyce-Davies in the edited text *Decolonizing the Academy: African Diaspora Studies* (2003) writes of the intention behind the academy being "the most colonized space" as the "site for the production and re-production of a variety of discourses which keep in place certain colonial structures which have as their intent the maintenance of Euro-American hegemonies at the level of thinking and therefore in the larger material world" (2003, ix). Similarly, Achille Mbembe has been clear that the academy is part of a larger system of global control which has "turned higher education into a marketable product bought and sold by standard units" (2015). Like Boyce-Davies and Mbembe, it is clear to me as a Black woman who sits within these academic spaces that both the structure and function of the academy remain securely planted within its racist, sexist capitalist legacies and that the "lingering racisms" within these academic geographic spaces impact how "race is theorized, lived, debated, departmentalized, inter-departmentalized, and mapped out in university settings" (McKittrick 2016, 4). We see this every day for example in the pivot to diversity initiatives which according to Anneth Kuar Hundle have been largely "shaped and retooled in the context of broader ideologies of liberal multiculturalism and neoliberal capitalist governance in this period" (2019, 292). Many Black feminists before me have written about their endurance of academic dispossession, marginalization, surveillance, and violent exploitation (Alexander 2005; Fries-Britt and Turner Kelly 2005; Harley 2008), which today has been made particularly acute by the simultaneous denial and disavowal of the historical impacts through university diversity statements, land acknowledgments, and the deployment of various equality working groups and committees. This dichotomy produces a particular set of university logics, rooted in the plantation, that are in fact in many ways quite illogical and in some ways absurd to Black folx.

These university illogics were made acutely apparent in 2020 and 2021 with the onset of the COVID-19 global pandemic and the anti-racist protest sparked in response to the murders of George Floyd and Breonna Taylor. The ways in which the university responded in supposed solidarity with those disproportionately negatively impacted directly, indirectly, physically, psychologically, and/or economically by the pandemic and the continuous fight for Black lives have been wholly inadequate, insufficient, disingenuous, and in some cases even patronizing, resulting in heightened tensions, conflicts, and increased academic displacement. These actions include for example the empty statements in support for Black lives, and the hiring of more Black people in diversity positions without power and with limited resources. Awino Okech and Shereen Essof for instance write about how they

> watched universities and organisations tripping over themselves to issue statements and make commitments to addressing institutional racism … [leading] to a flurry of new academic posts on Black Studies or Critical Race Studies … making claims about its commitments against structural racism [to demonstrate] that there was work being done rather than performative public claims.
>
> (2021, 3)

Okech and Essof are clear that it is empty gestures of inclusion and solidarity like these that

> fail to appreciate the way our economies are racialised, thus marginalising whole categories of people on the edges of educational institutions from primary school through to universities [which lead] to Black women academics drop[ping] out of the sector due to the isolation, hostility, and devaluation of their scholarship.
>
> (2021, 3)

When for example we look at the recent case of MacArthur genius, Pulitzer Prize award-winning journalist, and creator of the 1619 project (The New York Times 2019) Nikole Hannah-Jones who was recently initially denied tenure at the University of North Carolina (The New York Times 2021), we can certainly see the illogics of it all especially when no one in that position before her was ever denied tenure. But they were all white! In this situation, Hannah-Jones is an acclaimed scholar which meant that her case garnered national publicity and mobilized Black feminist and other communities (Anderson and Svrluga 2021). But what of the Black women academics who are not award-winning, public, or popular scholars who may be suffering because of plantation illogics in our contemporary period? How are they supported at this time? These questions need to really be considered particularly with the recent backlash against critical race theory (CRT) and the easy acceptance of and incorporation by some institutions of former US President Donald Trump's criticism of CRT (Lang 2020), as well as the current assault on CRT in states like Florida (Craven 2021) where academic institutions have yet to make any strong nationally coordinated statements in support of the field and its scholars, who are predominantly Black and Brown – all of this happening as Juneteenth has been made a federal holiday. It makes no sense at all – these logics are in fact illogical! What has been made very clear in this time of dislocation, dispossession, disease, and death during COVID-19 and Black uprisings, is the need for "fewer institutional statements in support of BLM, and greater commitment to making universities places where we as Black people, and Black women in particular, do not feel isolated and, therefore, experience them as hostile" (Okech and Essof 2021, 4). Quoting Okech and Essof again, universities also need to ensure for example that "when [we] walk into a meeting, [we are] not the sole Black person or Black woman, and always, therefore, confined to addressing race and gender issues because if you do not, no one will" (2021, 4).

The above demonstrates that even more during this time, "underlying institutional structures remain fundamentally unchanged [and] … diversity becomes a highly valuable commodity and a powerfully legitimizing institutional force" (Rosa and Bonilla 2017, 202). The experiences of Black women in academic spaces often end up with them struggling within systems of domination which produce what McKittrick refers to as "difficult entanglements of racial encounter" (2011, 949). Like the past, the contemporary conditions of academic struggle produced by these encounters – both racial and sexual – end up producing fugitive conditions for those who are the most marginalized within academic spaces. During the past year, many Black women have left academia temporarily or altogether. Many, including myself, have shifted their/our priorities away from serving solely in the academy and away from engaging academic illogics. These fugitive conditions have also materialized many possibilities as more Black women academics are pivoting toward working and being in community with activist organizations. As we search for our sense of place, we consider McKittrick's statement that "spaces of encounter … hold in them useful anti-colonial practices and narratives" (2011, 955) such that Black feminist fugitivity have the potential to disrupt "disciplined ways of knowing, and the spatial workings of knowledge" (McKittrick 2016, 4) that are produced through plantation illogics.

Fugitive Possibilities in Our Contemporary Times: The Example of Black Women Radicals

When we think of Black women who have left the academy or who have their feet both inside and outside of the academy, we cannot only think about what they are leaving behind or escaping but rather where they are going and what knowledges they are taking with them and communities they are building through their movements. In 2020 in my own fugitive movements, I started a collaboration with the Black feminist activist organization Black Women Radicals (BWR). BWR is a "Black feminist advocacy organization dedicated to uplifting Black women and gender expansive people's radical political activism. Rooted in intersectional and transnational feminisms and Womanisms" (Black Women Radicals 2022). At the invitation of BWR executive director and creator Dr. Jaimee Swift and Black feminist scholar and advocate Dr. Nana Afua Brantuo, I co-created a Caribbean feminist series and reading list (Black Women Radicals 2020) which featured the amazing activist, creative, and scholarly work of Black Caribbean women scholars/activists. The goal of this project was to not center our sense of Black feminist intellectual, creative, and artistic identity and space within the academy (Span and Sanya 2019), recognizing that part of our

> intellectual task … to work out how different kinds and types of voices relate to each other and open up unexpected and surprising ways to think about liberation, knowledge, justice, history, race, gender, narrative, and blackness … many divergent and different and relational voices of unfreedom are analytical and intellectual sites that can tell us something new about our academic concerns and our anticolonial futures.
>
> (McKittrick 2016, 5)

In doing so we can identify all types of Black feminist fugitives and un/re/cover their possibilities for Black feminist futures both within and beyond the academy.

Throughout 2020 we created what Carole Boyce-Davies theorizes as an elsewhere space, one that resonates deeply with me as a Black Caribbean woman, and which Boyce-Davies writes in her book *Black Women, Writing and Identity: Migrations of the Subject*, is a place more "of diasporic imaginings than the precisely locatable … oriented to articulating presences and histories across a variety of boundaries imposed by colonizers" (1994, 88). Quoting Boyce-Davies again, this time in her biography of the life of Claudia Jones, in this space we found "the kind of intellectual work produced organically outside of the academy [elsewhere] and accord that work the same weight and space one gives to academic production. … one has to undo the narrow equivalence of intellectual work with the academy" (2007, 10). In this space, we held virtual panel discussions where we engaged Afro-Caribbean feminist scholarship, activism, advocacy, issues facing Afro-Caribbean women, and celebrated the life and legacy of Guyanese scholar and activist Andaiye. Throughout the series, I was reminded over and over again of the legacy and livingness of Afro-Caribbean feminisms including how it has been in conversation with, impacted and has been impacted by US-centered Black feminist scholarship, advocacy and activism, such that as Patricia Hill Collins quoting Peggy Antrobus states, Black feminisms is not confined by the geographical borders of the US (Collins 2000). This acknowledgment is made while also paying attention to the ways in which impact does not always mean, according to V. Eudine Barriteau, direct engagement (2006). In this series, we attempted to enact gorgeous interaction with each other as Black women with our own socio-spatio-temporal peculiarities to realize the possibilities of our intellectual, advocacy, activist and artistic allegiances.

The series, which is still ongoing, has resulted in the building of Black Caribbean feminist friendships among each other, with our guests and with the hundreds of people who have attended our virtual events, and have started other new and exciting collaborations. What is more, we have found such fulfillment in the work and in the permission we have given ourselves to be curious and creative methodologically, to be in community in a time of isolation and share both intellectually and emotionally as Black feminist praxis. During this time, we, according to Saran and Haynes, embraced "our emotions as methodological tool" (2019, 1187). We were able to tap into deep emotions of anger, fear, sadness, and joy, allowing ourselves to feel the fullness of our experiences and in being alive, to "breathe life … into Black pain and recenter research within Black humanity" (2019, 1187). The space we created in our ongoing periods of exodus and flight from the academy throughout the pandemic allowed me to temporarily disengage with the illogics of the academy and the "mute allies, who don't know what to say or who think the demand at this moment [is] silence" (Okech and Essof 2021, 5). It allowed me the opportunity to engage in "real dialogue and listening" (Okech and Essof 2021, 5). It is the creation of a space on your own terms. As Black women, we are clear that in this virtual Caribbean feminist fugitive space, we do not accept prejudice in any from, and at the beginning of every series, Dr. Swift makes it clear that participants will be kicked out immediately for any displays of transphobia, racism, anti-blackness, misogyny, or misogynoir. In this space, far away from, "some dimly lit corridor … in the main auditorium" we illuminated each other, "derive joy in Black womxn reclaiming histories of knowing, of healing, centring music and dance, supporting learning and reflection in order to build collective power, maintain health, strength, energy, and razor sharp strategy" (Okech and Essof 2021, 6).

The fugitive space of the Caribbean feminist series has reminded me, in this time of multiple pandemics where Black life is seemingly tethered to trauma, death, and protest, just how intricately we are connected. And as I sit in this liminal spatiotemporality, preoccupied with survival and yet hopeful for the future, I learn from the examples of our foremothers of how we can continue to engage and sustain critical interventions, enacting fugitive possibilities toward present and future survival and thriving. In the final section of this chapter, I take seriously McKittrick's position that we can learn from those who came before us how to enact Black futures. She writes: "[s]ome black creative works, then, allow us to see into the future. I would argue, in fact, that the creative text that returns to the past intends to give blackness a future" (2016, 15). This is my intention.

Decolonial Futurity as Fugitive Possibility

According to Christina Sharpe, we, along with our ancestors, "inhabit knowledge that the Black body is the sign of immi/a/nent death" (2016, 71). We, like them, are often relegated to the status of "non-historical space-takers" (Cooper 2016). And yet, being branded with the mark of death and literally marked as out of time and space provides us with the fire that lights our desire for and belief in a more just future. The "not quite spaces of black women provide alternative paths through traditional geographies and take into account a political agenda concerned with racism-sexism, objectification, captivity, and respatialization" (McKittrick 2006, 62). Here I read McKittrick and Sharpe together as a way of demonstrating how knowing that our ancestors survived the longue durée of being rendered historical non-humans without place or space provides us with a knowledge, as Sharpe writes, that they are "with us" as we seek our own futurity. What Sharpe's annotation describes is a potent and affective connection which is able to sustain a deep and important sense of historical, political, and cultural memory, haunting and beauty, that highlights the agency and resistance efforts of Black people to secure our future.

Our ancestors, literally and figurative give/gave us life. They provided us with aspiration, that is "keeping and putting breath in the Black body" (Sharpe 2016, 130). As plantation illogics continue to pervade the contemporary university, I grapple with the question of how does the Black feminist transnational researcher situated within and yet beyond the university space in this time, and affected by these new iterations of plantation illogics, continue to pivot away from the frameworks of coloniality and neoliberalist knowledge regimes? How do I learn from my ancestors, pull from their memories to secure my own future, the futures of those I care about and those who will come after me? I posit that critically securing this future is the development of a decolonial knowledge project which is attentive to not only spatial but also temporal crossings such that I can reach back across time to the fugitive crossings of my Black ancestors to glean both situational, methodological, epistemological, and practical examples of fugitivity to imagine alternative destinations, both scholarly and otherwise to ensure our intellectual, that is, scholarly but also activist and artistic work, has a livable future and flourish.

As evidenced above, university-moves are premised on their investment in neoliberal capitalist and colonialist foundations, such that in simply inhabiting the space we will not get to a place of justice. To get to this space as a Black transnational scholar I practice a decolonial approach that is about "becoming and unraveling … [and] not about moving past or healing historical violences" (Vimalassery, Hu Pegues, and Goldstein 2016, p. 14). To achieve this unraveling, I engage with the aspects of plantation logics that are about imagining otherwise and other ways of the colonial legacy embedded within our society today and in which also remain the methods and strategies developed and used by those across the globe who have had to survive the plantation's historical brutality. These creative theoretical, ideological, ontological and practical methods include the ways Black people were able to make alternative scholarly and physical space through translating and sharing, network building, archiving and historicizing, and movement building, as they moved as fugitives in and out of oppressive spaces toward futurity.

Utilizing these methods which "move" beyond the staying in place required by the university, imagining the possibilities and futurity, invoking the ways in which historical communities conjured and enacted ancestral rituals to interrupt their geographical suturing, or more specifically, their capacity to imagine the idea of escape and mobility I think about how my own fugitive moves can work to guarantee a future for those who come after us. As a Black feminist transnational scholar, I acknowledge many ways in which I can enact movements, and in the remainder of the chapter I focus on four methodologies pulling from Black and transnational feminist legacies from which to draw on decolonial fugitive quests; these are methodologies of 1) the palimpsest, 2) Sankofa, 3) building networks of solidarity, and 4) maintaining traditional and cultural knowledge and care practices.

The transnational feminist researcher has always engaged with past knowledge as a type of palimpsestic work (Spivak 1988), looking beyond the constraints faced in the here and the now present. In a similar vein, Black feminists have also long engaged the concept of Sankofa, the Akan concept translated from Twi to English as to "go back and fetch what you forgot" (Temple 2010). Both of these methodological moves require that we not be tethered to a geographical time and space and by the relational dynamics of colonialism (Wilderson 2020, 228) but that we enact Black and transnational, often clandestine rituals of the past to gather present assemblage for an im/possible flight. Applying both the transnational and Black feminist concepts of the palimpsest and Sankofa provides an analytical opening that can guide our footsteps as we navigate the contemporary space of the academy which remains in place, that is, still implicated in the colonial plantation system used to justify the illogics of slavery, even as it portends to be moving forward in its rhetorical and performative commitments to diversity. The system of colonial transnational slavery which the academy was used to justify, which funded and built it, and which logics it continues to rely upon to secure its future, precipitated many ancestral fugitives like Harriet Tubman, whose legacy I invoke to demonstrate how fugitivity and futurity of Black people are always connected and hence why understanding the logics of the plantation is still important.

Looking back to Tubman's legacy allows me to understand the commitment it takes, both literally and figuratively, to be a fugitive; the contradictions and implications. Scholars have written extensively about Tubman's legacy, the creative and the scholarly fugitive praxis she developed to lead many out of colonial enslavement. For example, Vimalassery asks of Tubman who escaped the plantation in 1849 at the age of 27 and left her husband and extended family because of the danger of being resold (Hobson 2019),

> how did she convey meaning through changes in timbre, her choice of ornamentation, or by altering scales? What was the rhythm of her breath, or her phase? Each of these attributes is described in the archive of …her life and actions, but each of these, we can only imagine. Her distance from literacy is an immersion in a culture of orality and living embodiment, and the result, for us, is not so much a production of ignorance, as realms of knowledge that are irrecoverable through familiar empirical methods."
>
> (Vimalassery 2016, 4–5)

In addition, the fact that she left her family attends to the complications of staying in place academically versus escaping academic boundaries. Our fugitive movements are risky, and our desire to leave/get out/run/escape must be considered alongside knowing that in leaving we will leave others who we love but who cannot leave or want to stay behind. Tubman was a fugitive from the law as she defied conventional plantation logics. When we think of the fact that she and the others she guided to freedom were considered property, she was in fact stealing from the plantation. Her legacy gives us permission "to steal and repurpose the university's resources for collective study, acting as a 'fugitive network'" (Meyerhoff 2019, 17). Tubman's example also points us to those of Linda Brent/Harriet Jacobs and The Crafts (Snorton 2017) all of whom had to engage in various forms of creative concealments, "trickery" and disguise in their flight to a better future, but that entailed planning, courage and daring.

As a Black transnational feminist scholar from the Caribbean, it is also important that I address the legacy of interconnectedness of the plantations across geographical space to my own fugitive quests. Like Harney and Moten write, what comes with these movements for me is

> never being on the right side of the Atlantic is an unsettled feeling, the feeling of a thing that unsettles … a feeling, if you ride with it, that produces a certain distance from the settled, from those who determine themselves in space and time, who locate themselves in a determined history.
> (2013, 97)

But what does moving mean for our new worlds and intimacy if we leave love and kin behind? Historically, this means engaging the methodologies of building and having trusted solidarity networks, and a reliance on our cultural inheritances to inform our creative and intellectual interventions. Tubman, and Linda Brent/ Harriet Jacobs, among other fugitives, relied on those who could comfortably inhabit transparent space but who recognized them as human to form networks of allyship. We also learn from Tubman's example how we can use traditional creative knowledge as signposts; the patchwork of quilts and the North Star as our guides rather than traditional maps. Like McKittrick, we understand how mapping and the construction of transparent geographical space is socially constructed to produce conditions of non-belonging (2006). As such we travel with the knowledge produced in our mothers' gardens (Walker 1983), their kitchen tables, hair salons, music, and poems, from "spaces of knowledge production outside of those legitimized by the academe … essential in developing a way of knowing …beyond that of a racialized spectacle" (Johnson and Joseph-Salisbury 2018, 151). As Black feminist Patricia Hill Collins writes, we bring with us knowledge from "alternative institutional locations and among women who are not commonly perceived as intellectuals" (Collins 2000, 14). According to Cedric Robinson, these types of knowledges are part of the Black radical tradition that has been saving our lives. Robinson writes for example that centuries of resistance become apparent in "the syncopations and the phrases, the scamp and the beat, the lyric and melody of Black language, Black beliefs, Black music" (1983, 315).

These intellectual traditions are sometimes missing from the official archival knowledge repositories in which Black women show up only through the ways in which they encounter power (Hartman 1997). As such, and as Black feminist scholars, we must therefore appreciate the multiple creative, oral, cultural, intellectual traditions, and geographic locations we have and continue to look for and draw therefrom in order to plot our future. I look to the examples of the fugitives from the Caribbean plantation, long referred to as maroons, the legend of Nanny of the Maroons and maritime marronage, ancestors who not only fled, and fought, but created spaces of freedom that persist even today. Their creativity and audacity are always present with me as I navigate the knowledge production regime situated literally and figuratively as a foreigner. These ancestors are what Edouard Glissant refers to as "ecological actors who cannot be bound by the history and artifacts of slavery" (Allewaert 2008, 355). I invoke the audacious legacy of Mary Prince for example, the first woman to present a petition to the British parliament. Prince's story is one of an enslaved woman born in Brackish Pond, Bermuda, in 1788, who was able to navigate transnational geographic, ideological and political borders in order to obtain and retain her freedom. Prince was able to take ownership of her story and dictate it to Susanna Moodie, a writer and member of the London anti-slavery movement. Her story of resistance, *The History of Mary Prince: A West Indian Slave*, was published in 1831 in which Prince spoke back to and rebuked her former enslavers. Prince's fugitive movements allowed her to speak back to a colonial system that brutalized her, and her audaciousness moved her to sue for libel when those in the pro-slavery movement attacked the veracity of her published account of her own life. I look at fugitives like Prince as I create my subversions. Like Prince, who engaged directly with systems that can offer recourse to her plight but also with fugitive paths, I seek to be a

> subversive intellectual ... [who] engage[s] both the university and fugitivity ... neither trying to extend the university nor ... a room of ... [my] own, [I] want ... to be in the world, in the world with others and making the world anew.
>
> (Halberstam 2013, 9–10)

Finally, as academic fugitives, while knowing that our movement will in fact save our lives, and can guarantee the lives of those who come after us, we must be wary that as we move we do not deplete ourselves to the point of death. I acknowledge that fugitivity is difficult, fugitives do not always get away with their lives, and risk recapture, take the example of Margaret Garner. Even in instances where fugitives escape with their lives they risk both mental and physical disfigurement – as with Dana in *Kindred* (Butler 1979). Running for freedom can also mean running toward Black death, such that fugitivity might put in motion death because Black life is always tethered to Black death such that to use the concept of Sankofa as we do, needs to make theoretical space for freedom that means jumping overboard into death, that is sacrifice of self/ career/ for a greater sense of freedom as our spirits return to our homeplace. In thinking about homeplace, I bear in mind the words of bell hooks who states, "those who dominate and oppress us benefit most when we have ... nothing left, no "homeplace" where we can recover ourselves" and care for our own (hooks 1990, 43). As such, like Christina Sharpe, I

> want to think 'care' as a problem of thought ... in the wake as a problem of thinking and for Black non/being in the world. Put another way, ... [this] work that insists and performs that thinking needs care ('all thought is Black thought') and that thinking and care need to stay in the wake.
>
> (2016, 5)

In so doing I pay attention to the ways I can enact and engage a method and praxis of care as I move and encounter the whir and the psychosocial vertigo of being stretched across time and space. I engage Alexis Pauline Gumbs' work, as she commemorates Tubman's legacy through retreat and pilgrimage from the enslaved uprising to the subsequent Black lesbian group, the Combahee River Collective (Gumbs, 2014) whose words of doing this work because they love themselves, is what drives my moving with and through our ancestors such that I "no longer have to leave our ancestors at the door; they not only speak through, but they teach through us. And that is the true essence of how humanity survives, thrives, and becomes" (Morton 2016, 35). Care in this way is, according to Jennifer C. Nash,

> a practice for tending to what has already been lost and what might be lost, a political tool for maintenance of self and collective, that is always oppositional to the logic of the state… It is thus a practice of being deeply attuned to historical and ongoing violence, and to loving in the midst of it, a strategy of examining black life and the structures that seek to constrain that life.
>
> (2019, 79)

What I hope to draw attention to in this section of the chapter is the ways in which we can pull from fugitive examples of creative knowledge production that draw from traditional and cultural knowledge, kin keeping, care and solidarity networks to secure our futurity within and beyond the academy. While human life according to McKittrick, "is marked by a racial economy of knowledge that conceals – but does not necessarily expunge – relational possibilities" we can rest assured that plantation fugitives have used the logic of these systems to circumvent and "construct a reality that is produced outside, or push[es] against, the laws of captivity" (2015, 8). As the beneficiaries of these creative and decolonial acts we are entrusted with ensuring that these fugitive knowledges continue to find safe passage through us to future generations. I use these examples as guides for how I can decide to engage or disengage with current iterations of university plantation illogics, how I can find retreat in the stories of sacred Black spaces, legends, traditions, consciousness, rebellions, rituals, practices, communities and resistances (Robinson 1983, 310). These past examples of the possibilities of fugitivity provide us in our present endurance of this iteration of the plantation with concrete ways in which, as Black transnational feminist scholars, we can navigate the academy through relying on these works to help us build an archive now for future generations. I engage in the transnational feminist palimpsestic method and the Black feminist use of Sankofa because we know according to Audre Lorde that there is a world we want to live in which is not attained lightly and for which we need to "begin talking, thinking, feeling ourselves for its shape" so that we do not "condemn ourselves and our children to a repetition of corruption and error" (in Byrd, Cole, and Guy-Sheftall 2009, 206).

Conclusion

I end this chapter with a quote from Christina Sharpe, who writes,

> Living as I have argued we do in the wake of slavery, in spaces where we were never meant to survive, or have been punished for surviving and for daring to claim or make spaces for something like freedom, we yet reimagine and transform spaces for and practices of an ethics of care (as in repair, maintenance, attention), an ethics of seeing, and of *being* in the wake as consciousness; as a way of remembering and observance that started with the door of no return, continued in the hold of the ship and on the shore. … This is an account counter to the violence of abstraction … An account of *care* as shared risk between and among the Black trans*asterisked.
>
> (2016, 130–131)

In this chapter, I engage with the notion of fugitivity as "separate from settling" (Halberstam 2013) by connecting our present plight with university plantation illogics palimsestically with the refusals of our fugitive ancestors. I engage with their memory to enact what M. Jacqui Alexander calls "as a Sacred dimension of self" (2005, 14). That is, their actions provide me with the sacred knowledge I can use daily to guide my fugitive crossings. I understand that remembering their legacy situates me and other Black women and our present "within a larger critical community – carrying part of the 'force of history with us'" (Alexander 2005, 152).

I also examine how the legacy and the logics of the plantation as it is has moved through time to form the contemporary university in 2020/21, has in the face of the COVID-19 pandemic and the anti-racist uprising, responded by falling back on its colonial underpinnings. It has been interesting to watch the ways in which the university was forced to rethink and redo how it has operated for centuries. Using plantation logics, I have argued that in being forced to strategize and respond to COVID-19 and the racial uprisings, the university relied on the strategies of the plantation and on a rhetorical commitment to keeping the knowledge production system founded in a colonial past in place as its strategy for moving forward. The impacts of the universities moving – spatially (online) and rhetorically (in defense of black lives) is in fact also a staying in place. There is no doubt that its responses to the events of 2020/21 is bound up in its colonial underpinnings and as such has implications for the futurity of Black bodies who inhabit these spaces. In examining the university as part of what Lisa Lowe calls "[t]he operations that pronounce colonial divisions of humanity … imbricated processes, not sequential events … ongoing and continuous in our contemporary moment, not temporally distinct now as yet conclude" (2015, 7), I posited that we must be wary of any university "moves" viewing them through a lens that accounts for "colonialism's multiple, uneven yet interlocking, violences through an extended contemplation that does not forsake contradiction and complexity for facile and overdetermined paradigms" (Vimalassery, Hu Pegues, and Goldstein 2016, 15). The strategies and logics of the plantation are still very much alive and well in the university which continues to work to ensure its own economic futurity at the expense of the futurity for Black people in the academy. Contemporary university-moves are grounded in "facades of inclusion without any change to dominant and unequal power structures or knowledge bases, and this is often the crux of the matter" (Sultana 2019, 35). The results are that these responses/movements are actually in the interests of a particular type of enforced habitation – a staying in place that necessitates increased surveillance and silence.

According to Grace Kyungwon Hong our work cannot "concede the future to the present, but imagines it as something still in the balance, something that can be fought over, … the work … of imagining, is revolutionary" (2008, 108). Like our fugitive ancestors, we cannot gamble our futurity upon the contemporary inner workings of the plantation and its iterations including the academy. Rather, we place it lovingly in the capable hands of our ancestors whose fugitive examples provide guidance on how to inhabit their examples of moving beyond the university in creative scholarly and artistic ways. We look to these examples as "field sites and require ethnographic attention through reflexive methodologies … establishing space for alternative presents and futures in the university – new cultures, philosophies, and epistemologies of citizenship, community, and difference that will engender the redefinition of the university itself" (Hundle 2019, 318). The strategic moves I outline in this chapter tackle some of the ways in which as scholars our spatiotemporal movements within and beyond the academy need to engage not only how we will get through this present performative moving of the academy that allows it to stay solidly in place, but also requires that as we make our encounter with these moves we ask ourselves "what futures might it open up" (Ahmed 2000, 145)? In recognizing how the ways in which we arrived in the academy is largely through the creative scholarly, artistic, activist, audacious intellectual work originating outside of and beyond the academy, within rich cultures, communities and relations, we realize the potential we hold as stewards of this knowledge to facilitate its safe journey into the future.[2]

Notes

1 It is important to note here that this chapter is reflexive and hence focuses specifically on my knowledge about and experiences within the structures and with initiatives at Predominantly White Institutions (PWIs) in the US. I acknowledge however, that the history of Historically Black Colleges and Universities (HBCUs) in many instances causes them to grapple with similar issues as PWIs. While the role of HBCUs has been to offer educational opportunities for African Americans (Allen & Jewel, 2002; Holmes, Land, & Hinton-Hudson, 2007), according to Esnard and Cobb-Roberts (2018) one must consider HBCUs within the context of their founding, some during a period of rampant direct colonization "by White missionaries espousing Christian ideals" and religious conservatism (Esnard & Cobb-Roberts, 2018, 403). There are still very audible and procedural resonances of these white, male, and Christian voices at HBCUs, some racist and sometimes implemented by those who share the racial identities of the majority student body population at HBCUs. The result is the sustaining of spaces where decolonial and anti-racist scholarship and praxes are confronted with dominant historical positions of power as many of them also depend on government funding. As such HBCUs'

relationships to power impact their mandate and the "organizational cultures and practices that are enacted within that space, … such practices can … [impact] issues of inclusion/exclusion and progression/regression of both students and staff within that environment" (Esnard & Cobb-Roberts, 2018, 404) such that they often "perpetuated the same practices that they are supposed to be standing against'" (378–9). HBCUs can be particularly violent and unhealthy places of Black women especially, and can result in emotional, psychological and physical distress from which they might seek flight.

2 Many thanks to Dr. Nana Afua Branto for her help in formulating my thoughts and some of the concepts developed in this chapter.

References

Ahmed, Sara. 2000. *Strange Encounters: Embodies Others in Post-Coloniality*. New York: Routledge.

Alexander, M. Jacqui. 2005. *Pedagogies of Crossing: Meditations on Feminism, Sexual Politics, Memory, and the Sacred*. Durham: Duke University Press.

Allen, W. R., and J. O. Jewel 2002. "A Backward Glance Forward: Past, Present, Future Perspectives on Historically Black Colleges and Universities." *The Review of Higher Education*, 25(3): 241–61.

Allewaert, M. 2008. "Swamp Sublime: Ecologies of Resistance in the American Plantation Zone." *PMLA*, 123(2): 340–57.

Anderson, Nick, and Susan Svrluga. 2021. "Pressure builds on UNC board to grant tenure to journalist Nikole Hannah-Jones." *The Washington Post*. Accessed 11 March 2022. https://www.washingtonpost.com/education/2021/05/26/unc-trustees-hannah-jones-tenure-1619/.

Barriteau, Violet Eudine. 2006. "The Relevance of Black Feminist Scholarship: A Caribbean Perspective." *Feminist Africa*, (7): 9–31.

Black women Radicals. 2020. Accessed 11 March 2022. https://www.blackwomenradicals.com/blog-feed/caribbean-feminisms-a-reading-list

Black Women Radicals. Accessed 11 March 2022. https://www.blackwomenradicals.com/.

Boyce-Davies, Carole. 2007. *Left of Karl Marx: The Political Life of Black Communist Claudia Jones*. Durham: Duke University Press.

———. 2003. "Introduction: Decolonizing the Academy/Advancing the Process." In *Decolonizing the Academy: African Diaspora Studies*, eds. Carole Boyce-Davies, Meredith Gadsby, Charles Peterson, and Henrietta Williams, ix–xvi. Trenton: Africa World Press Inc.

———. 1994. *Black Women, Writing and Identity: Migrations of the Subject*. New York: Routledge.

Butler, Octavia. 1979. *Kindred*. Boston: Beacon Press.

Byrd, R. P., Johnnetta Betch Cole, and Beverly Guy-Sheftall. 2009. *I am Your Sister: Collected and Unpublished Writings of Audre Lorde*. Oxford: Oxford University Press.

Campt, Tina. 2014. *Black Feminist Futures and the Practice of Fugitivity*. https://bcrw.barnard.edu/videos/tina-campt-black-feminist-futures-and-the-practice-of-fugitivity/

Collins, Patricia Hill. 2000. *Black Feminist Thought: Knowledge, Consciousness, and the Politics of Empowerment*. New York: Routledge.

Cooper, Brittney. 2016. The Racial Politics of Time, TED Talk. Accessed 11 March 2022. https://www.ted.com/talks/brittney_cooper_the_racial_politics_of_time?language=en

Craven, Julia. 2021. What Florida's "Critical Race Theory" Ban Tells Us About Anti-Antiracism. *Slate*. Accessed 11 March 2022. https://slate.com/news-and-politics/2021/06/florida-critical-race-theory-ban-what-it-tells-us-about-anti-anti-racism.html

Esnard, T., and D. Cobb-Roberts 2018. *Black Women, Academe, and The Tenure Process in The United States and Caribbean*. New York: Palgrave Macmillian.

Fries-Britt, Sharon, and Bridget Turner Kelly. 2005. "Retaining Each Other: Narratives of Two African American Women in the Academy." *The Urban Review*, 37(3): 221–42.

Gilmore, Ruth Wilson. 2022. *Abolition Geography: Essays Towards Liberation*. New York: Verso.

Gumbs, Alexis Pauline. 2014. "Prophecy in the Present Tense: Harriet Tubman, the Combahee Pilgrimage, and Dreams Coming True." *Meridians: Feminism, Race, Transnationalism*, 12(2): 142–52.

Halberstam, Jack. 2013. "The Wild Beyond With and For the Undercommons." In *The Undercommons: Fugitive Planning & Black Study*, eds. Stefano Harney and Fred Moten, 5–12. Brooklyn: Minor Compositions.

Harley, D. A. 2008. "Maids of Academe: African American Women Faculty at Predominately White Institutions" *Journal of African American Studies*, 12: 19–36.

Harman, Saidiya. 1997. *Scenes of Subjection: Terror, Slavery, and Self-Making in Nineteenth-Century America*. Oxford: Oxford University Press.

Harney, Stefano, and Fred Moten. 2013. *The Undercommons: Fugitive Planning & Black Study*. Brooklyn: Minor Compositions.

Hobson, Janell. 2019. "Of "Sound" and "Unsound" Body and Mind: Reconfiguring the Heroic Portrait of Harriet Tubman." *Frontiers: A Journal of Women Studies*, 40(2): 193–218.

Holmes, S.L., L. D. Land, and V. D. Hinton-Hudson 2007. Race Still Matters: Considerations for Mentoring Black Women in Academe. *Negro Educational Review*, 58(1–2): 105–29.

Hong, Grace K. 2008. ""The Future of Our Worlds": Black Feminism and the Politics of Knowledge in the University under Globalization. *Meridians: Feminism, Race, Transnationalism*, 8(2): 95–115.

Hooks, B. 1990. *Yearning: Race, Gender, and Cultural Politics*. Boston, MA: South End Press.

Hundle, Anneeth Kaur. 2019. "Decolonizing Diversity: Transnational Politics of Minority Racial Difference." *Public Culture* 3(12): 289–322.

Johnson, Azeezat, and Remi Joseph-Salisbury. 2018. "'Are You Supposed to be in Here?': Racial Microaggressions and Knowledge Production in Higher Education." In *Dismantling Race In Higher Education: Racism, Whiteness and Decolonising the Academy*, eds. Jason Arday and Heidi Sadia Mirza, 143–60. London: Palgrave McMillian.

Lang, Cady. 2020. President Trump Has Attacked Critical Race Theory. Here's What to Know About the Intellectual Movement. *Time*. Accessed 11 March 2022. https://time.com/5891138/critical-race-theory-explained/

Lowe, Lisa. 2015. *The Intimacies of Four Continents*. Durham: Duke University.

McKittrick, Katherine. 2016. "Diachronic loops/deadweight tonnage/bad made measure." *Cultural Geographies*, 23(1): 3–18.

McKittrick, Katherine. (Ed.). 2015. *Sylvia Wynter: On Being Human As Praxis*. Durham: Duke University Press.

———. (Ed.). 2013. "Plantation Futures." *Small Axe* 17(3): 1–15.

———. (Ed.). 2011. "On Plantations, Prisons, and a Black Sense of Place." *Social & Cultural Geography*, 12(8): 947–63.
———. (Ed.). 2006. *Demonic Grounds: Black Women and the Cartographies of Struggle*. Minneapolis: Minnesota Press.
Mbembe, Achille. 2015. *Decolonizing Knowledge and the Question of the Archive*. Africa is a Country. https://africaisacountry.atavist.com/decolonizing-knowledge-and-the-question-of-the-archive
Meyerhoff, Eli. 2019. *Beyond Education: Radical Studying for Another World*. Minneapolis: University of Minnesota Press.
Moten, Fred. 2018. *Stolen Life: Consent Not to Be A Single Being*. Durham: Duke University Press.
Morton, Berlisha R. 2016. Ain't Nothin' Wrong with Cleanin' Houses: Utterance on Southern Womanism and the Search for Our Mamas' Gardens. In *Race, Gender, and Curriculum Theorizing: Working in Womanish Ways*, eds. Denise Taliaferro Baszile, Kirsten T. Edwards, and Nichole A. Guillory, 17–36. Lanham, MD: Lexington Books.
Nash, Jennifer C. 2019. *Black Feminism Reimagined After Intersectionality*. Durham: Duke University Press.
Okech, Awino, and Shereen Essof. 2021. "Global Movement for Black Lives: A Conversation Between Awino Okech and Shereen Essof." *European Journal of Women's Studies*, 00(0): 1–9.
Robinson, Cedric J. 1983. *Black Marxism: The Black Radical Tradition*. Chapel Hill: University of North Carolina Press.
Rosa, Jonathan, and Yarimar Bonilla. 2017. "Deprovincializing Trump, Decolonizing Diversity, and Unsettling Anthropology." *American Ethnologist*, 44(2): 201–8.
Saran, Stewart, and Chayla Haynes. 2019. "Black Liberation Research: Qualitative Methodological Considerations." *International Journal of Qualitative Studies in Education*, 32(9): 1183–9.
Sharpe, Christina. 2016. *In the Wake: Of Blackness and Being*. Durham: Duke University Press.
Snorton, S. Riley. 2017. *Black on Both Sides: A Radical History of Trans Identity*. Minneapolis. University of Minnesota Press.
Span, Christopher M., and Brenda N. Sanya. 2019. "Education and the African Diaspora." In *The Oxford Handbook of the History of Education*, eds. John L. Rury and Eileen H. Tamura, 399–411. Oxford: Oxford University Press.
Spivak, Gayatri C. 1988. "'Can the Subaltern Speak?'" In *Marxism and the Interpretation of Culture*, eds. C. Nelson and L. Grossberg, 271–313. Champaign, IL: University of Illinois Press.
Sultana, Farhana. 2019. "Decolonizing Development Education and the Pursuit of Social Justice." *Human Geography*, 12(3): 31–46.
Temple, Christel N. 2010. "The Emergence of Sankofa Practice in the United States: A Modern History." *Journal of Black Studies*, 41(1): 127–50. https://doi.org/10.1177/0021934709332464
The New York Times. 2019. https://www.nytimes.com/interactive/2019/08/14/magazine/1619-america-slavery.html
———. 2021. https://www.nytimes.com/2021/05/19/business/media/nikole-hannah-jones-unc.htmlThe
Vimalassery, Manu. 2016. "Fugitive Decolonization." *Theory & Event* 19(4).
Vimalassery, Manu, Pegues, Hu Juliana, and Goldstein, Alyosha. 2016. "Introduction: On Colonial Unknowing." *Theory & Event*, 19(4).

Walker, Alice. 1983. *In Search of Our Mothers' Gardens: Womanist Prose*. Harcourt.
Wilderson, Frank B. 2020. *Afropessimism*. New York, NY: Liveright Publishing Corporation.

Afterword

Julia S. Jordan-Zachery

As a little girl, if I went to my gran and complained about some things, such as the humid August summers in the concrete jungle of Brooklyn, NY, she would cheekily reply, "Nobody didn't call yuh." I would chuckle and find ways to manage the sweat that pooled in various parts as I tried to get lost in the latest book I'd borrowed from the public library. Books allowed me to find refuge in places—real and imagined—as I tried to figure out what it meant to be an "immigrant," particularly a Black, young woman immigrant from the Caribbean.

Years later, as I faced ongoing race-gender violences, I would reflect on my gran's response, "Nobody didn't call yuh," as I pondered how am I here (Jordan-Zacherry, 2019). Sitting in darkness, physical and emotional, I reflected on what it meant to be this little girl from Barbados who became the Black woman academic who faced persistent race-gender violences. And like the teenager who had migrated to the U.S., who found solace in reading, I turned to text to help me make sense of my experiences. But more importantly, I turned to Black feminist texts (whether identified as such by the author or just by me). These texts helped me to enter a deep period of self-making by affording me the space I needed to enter my interior. Eventually, I would write about the Black feminine interiority and the Black woman's feeling body (Jordan-Zachery, 2022). I write, "we [Black women] access [our] truth living by going into the interior, experiencing our feelings, and taking the knowledge these feelings offer to come into the world—a world that sometimes tries to violently expel Black women." As I further argue,

> Black women use feelings, accessed by going into the interior, to engage in the process of self-actualization. As such, this is a book on Black women's interiority and the knowledge that rests there, and how they use this knowledge to see themselves to make themselves real.

How do Black women make themselves real in the context of transnationalism, globalization, capitalism, anti-Black womanness, and the list goes on?

Global Black Feminisms helps us explore and better understand how Black women make themselves real—we go into the interior and come out in community in our quest for justice. But let me come back to my readings as I tried

DOI: 10.4324/9781003143550-17

to grapple with "Nobody didn't call yuh." As part of my journey of going into the interior, I read Dionne Brand's short story "Blossom: Priestess of Oya, Goddess of Winds, Storms and Waterfall" in *Sans Souci and Other Stories*. I found myself captivated by Blossom, the central character. And as I read *Global Black Feminisms*, I find myself dwelling on Blossom yet again. Blossom seems to embody much of what this insightful and thought-provoking text is doing. In this short story, we get to see how a Black woman makes herself real when she realizes that "Nobody didn't call yuh," but society has it as such that you are here. Blossom shows us what it means for a Black transnational woman to go to the interior to engage in the process of self-actualization and to come back out to challenge the logics and techniques of oppressive structures.

Through her experiences with race-gender violences—including migration, white supremacy, patriarchy, etc., Blossom has a moment where she goes into her interior. Some fashion this as a mental breakdown (Garvey 2003). I imagine it as her journey to tap into her feeling body. And during this journey, she encounters Priestess Oya—the Yoruba Orisha who is associated with winds, storms, and waterfalls. Judith Gleason (1987, 1) characterizes Oya as:

> the leader of the market women in Yoruba communities…offer[ing] special protection and encouragement in negotiation with civil authorities and arbitration disputes. Thus, one may speak of Oya as patron of feminine leadership, of persuasive charm reinforced by àjé—an efficacious gift usually translated as "witchcraft."

Brand ([1988] 1989) writes,

> Next thing Blossom know, she running Victor down Vaughan Road screaming and waving the bread knife. She hear somebody screaming loud, loud. At first, she didn't know who it is, and is then she realize that the scream was coming from she and she couldn't stop it. She dress in she nightie alone and screaming in the middle of the road…She wake up the next morning, feeling shaky and something like spiritual. She was frightened, in case the crying come back again…She had the feeling that she was holding she body around she heart, holding sheself together, tight, tight. She get dressed and went to the Pentecostal Church where she get married and sit there till evening. For two weeks this is all Blossom do…During these weeks she could drink nothing but water.
>
> (38)

She drank water, sat with her feelings, and allowed Oya to guide her deeper into her articulation of self. Through her sitting with Oya, Blossom gained "the power to see and the power to fight; […] the power to feel pain and the power to heal" (Brand [1988] 1989, 40). This is the beauty of what it means to be Black and woman—what so many refer to as #Blackgirlmagic. The magic is what happens when we are able to be in relation first to self and then to others (in the case of Blossom, it is Oya, and in the case of those who contribute to this text, it is the teacher/student). Through her relationship with Oya, Blossom engages in a practice of self-naming while she encounters Black and Black feminine suffering resulting from system failures that produce race-gender violences and trauma. "Because [Black women] were/are products of separations and dislocations and dismemberings, people of African descent in the Americas historically have sought reconnection" (Boyce Davies 1994, 17). When we can enter the interior to sit in community with the ancestors, deities, and others, we can engage in the hard work of reconnection that Boyce Davies describes. And we see this with Blossom and through the chapters of this text—the process of reconnection that results in self-naming and self-actualization.

Blossom serves as an archive for that pain and as a map for how to embody agency. Blossom, like the various authors of this text, embodies desire, loss, and longing for a different future by dipping into the past to bring it to the present so that she/they/us can birth a different future. She subverts hegemonic logics and techniques of power to manifest her "Speak Easy." Blossom's "Speak Easy" enterprise allows others to stop by to drink, fellowship with others, to witness her when Oya emerges in her body, and to offer them healing that is beyond religion and capitalist structures. Like Blossom, *Global Black Feminisms* affords us an opportunity to rest, to go into our interior to re(vision) the world using our identities as simultaneously Black and woman (among other identity markers). This book is another form of the "speak easy."

References

Boyce Davies, C. 1994. *Black Women, Writing and Identity: Migrations of the Subject.* London: Routledge.

Brand, D. (1988) 1989. "Blossom: Priestess of Oya, Goddess of Winds, Storms, and Waterfalls." In *Sans Souci and Other Stories*. Ithaca, New York: Firebrand Books.

Garvey, J. X. K. 2003. "'The Place She Miss': Exile, Memory, and Resistance in Dionne Brand's Fiction." *Callaloo*, 26(2): 486–503

Gleason, J. 1987. *Oya: In Praise of the Goddess*. Boston: Shambhala Publication.

Jordan-Zachery, J. S. 2022. *Erotic Testimonies: Black Women Daring to be Wild and Free*. New York: Albany, NY: SUNY Press.

——— 2019. Licking Salt: A Black Woman's Tale of Betrayal, Adversity, and Survival. *Feminist Formations*, 31(1): 67–84.

Index

Note: Pages in *italics* refer to figures and pages followed by "n" refer to notes.